Edited by

Barbara Marinacci

Introduction by

Linus Pauling

———

A Touchstone Book

Published by Simon & Schuster

New York London Toronto Sydney Tokyo Singapore

Linus Pauling

in His Own Words

Selected Writings,

Speeches, and

Interviews

TOUCHSTONE
Rockefeller Center
1230 Avenue of the Americas
New York, NY 10020

SIMON & SCHUSTER and colophon are registered trademarks
of Simon & Schuster Inc.

TOUCHSTONE and colophon are registered trademarks
of Simon & Schuster Inc.

Designed by Karolina Harris

Manufactured in the United States of America

1   3   5   7   9   10   8   6   4   2

Library of Congress Cataloging-in-Publication Data
Pauling, Linus, 1901–
Linus Pauling in his own words : selections from his writings,
speeches, and interviews / edited by Barbara Marinacci ;
introduction by Linus Pauling.
p.   cm.
"A Touchstone book."
Includes bibliographical references and index.
1. Pauling, Linus, 1901–   .
2. Science—History.   3. Scientists—United States—Biography.
I. Marinacci, Barbara.   II. Title.
Q143.P25A3   1995
081—dc20   95-31123
CIP
ISBN 0-684-80749-1 (Hardcover)
0-684-81387-4 (Pbk.)

All permissions are listed on pages 306–9.

To the memory of Linus Pauling,

which will be ever vivid and admiring,

and to the long and flourishing life of the

nonprofit organization he founded to

conduct research and education on the

connection between human health and nutrition—

THE LINUS PAULING INSTITUTE OF SCIENCE AND MEDICINE

Contents

———

IV   Nutritional Medicine *1954–1994*

Introduction

by LINUS PAULING

I was born at the beginning of this century, in 1901, and am living close to its conclusion—a fine period to have been a scientist. No century before this one has experienced such changes in the way we humans live, do our work, or even think. Science is responsible for most of these changes.

In the later years of life, when people usually write their own histories, I have been so involved with scientific interests and with issues relating to the well-being of humankind that I did not spend time in telling my own life story for book publication. I am now ninety-three years of age and still have other work to accomplish. So this book will take me as close to writing my memoirs or autobiography as I shall ever get.

It happens that in various writings, in talks, and in interviews, I have often reminisced about such matters as my childhood and early education, my adventures in different fields of science—physical, life, and medical—my acquaintance with other scientists, my involvement with the peace movement, and my family life. Frequently, too, I have expressed my opinions on a great many other topics. I know that whole books might be made of my own experiences and reflections, told through words I wrote or spoke at different stages of my life. There are ample verbal materials for a historical record of my passage through almost the entire twentieth century. This book contains but a fraction of them.

For seventy-five years I have been writing articles and books. The first article, published in 1920 when I was nineteen years old, was my essay "The Manufacture of Cement in Oregon"; I was then an undergraduate student in chemical engineering at Oregon Agricultural College in Corvallis—now Oregon State University. Since then I have written many articles, lectures, speeches, and books, along with letters and reports. Only a portion of these have been published.

Through the years, busy with more compelling matters, I did not always collect what I wrote. Fortunately for me, however, I had help from other people—my wife, a succession of excellent secretarial and scientific assistants, and other professionals who wanted to keep track of my words. So it was possible for both the Linus Pauling Institute of Science and Medicine (LPI) in Palo Alto, California, and the Ava Helen and Linus Pauling Papers at Kerr Library, Oregon State University (OSU), to gather and preserve much of what I have written or said through the years. Additionally, the California Institute of Technology (CIT) in Pasadena has holdings relevant to my long stay there, from 1922 to 1964, and subsequent communications.

Two bibliographies of my published work have been prepared. The first, *Scientific Publications of Linus Pauling*, was done by Gustav Albrecht of CIT in 1968, and as its title states, it was concerned only with my scientific papers and books. It listed 370 papers and nine books. The second, assembled by my longtime assistants at the Linus Pauling Institute, Dr. Zelek Herman and Dorothy Bruce Munro, is periodically updated with new publications, and also revised as previously undiscovered publications of mine are brought to their attention. The master list at this time, which includes Albrecht's original list, has 1,055 entries but is growing. The Pauling Papers at OSU has a catalog detailing its collection. I was astonished to learn that many of these documents at OSU, put into digitalized form, will be available electronically in the near future to scholars everywhere in the world.

The entries in a complete list of my publications must be carefully weighed as to their importance or identity. Some entries are for sizable volumes, while others may be for brief letters printed in newspapers or for interviews that appeared in books or magazines. There are even a few publications from organizations such as churches or political groups that were not formally published but were machine duplicated from typewritten (not printed) texts. I should also point out that not all of the titles listed are separate items. Among the entries are translations of my books and scientific papers, as well as revised and updated book editions. Also, some articles were reprinted elsewhere, perhaps in a slightly altered form and under a different title, or included in an anthology.

Moreover, I am not always the sole author of the publications listed. Throughout my life I have had many scientific collaborators; the names of the principal ones are alongside mine in our publications. Occasionally, too, I have had uncredited assistance when composing an article or a letter to be published in my name, particularly for popular readership. Because my name is widely recognized, my opinions are frequently sought. Sometimes, if my time for writing is limited, I may work with a writer whose ideas and language are close to my own.

My life has been closely examined as well as my words. Thus far there have been several books published about me; one was a handsome pictorial volume printed in Japan, another a biography for young adults (by Florence Meiman White), still another a book (by Anthony Serafini) that contained many errors and misinterpretations of my life and work. At least several more biographies are in preparation, including those by Thomas Hager of Eugene, Oregon, and by Professor Robert J. Paradowski of the Rochester Institute of Technology. Both of these writers have interviewed me extensively. Additionally, short biographies or profiles have been published during the past half century of my life. Also, several television documentaries about me have appeared.

Up to now, however, there have been no general anthologies of my written words to provide lay readers with an overview of the many different interests that I have pursued during my lifetime. This volume selects published writings, lectures, and interviews. I understand that it will also contain some unpublished speeches, articles, and notes. The pieces here are chosen to represent the main routes and some byways that my investigations have taken me through in the world I have lived in for ninety-three years. They should also provide some of my thoughts about such matters as the creative process of discovery in science, ethics, human health, and the future of life upon this planet.

Like many scientists, I have always taken great joy in discovering as much as I can about the detailed workings of the world that we human beings inhabit. My own initial quest as a scientist was to explore the structure of matter, along with the behavior of particles of energy associated with matter's many forms. I started my studies of chemistry in the realm of inorganic materials; later I began to move into organic chemistry, which investigates carbon-containing molecules, some of which are or have been living matter.

It was not such a jump, then, for me to become involved in biochemistry. My research was connected with the structure and behavior of important substances within humans. The proteins especially interested me. The more I studied issues about health, the more I looked around me at a world populated by humans who were in perpetual deadly conflict with one another, for no good reason. I could not understand how this higher primate could assign to its modern form the species name with a double dose of the Latin word for "wise": *Homo sapiens sapiens*.

And so, as a lifelong advocate of democracy, I began to speak out. I thought that people trained to think as scientists might make valuable contributions in searching for solutions to problems in society and in shaping the human future in a positive way. I came to believe that a science of morality is possible—indeed, that it had become essential to achieve. The two Nobel Prizes that I received, in Chemistry (1954) and for Peace (1962), have demonstrated, I think, that one man in his life-time can be recognized for accomplishments in both science and hu-manism. I am equally proud of both endeavors.

I have not regretted my peace activism, although this has damaged my reputation as a scientist among certain people and institutions. Nor have I regretted giving my attention during the last quarter century to the application of a new theory of medicine, involving nutrition, to human well-being. Yet these efforts in addressing my concerns over the health and safety of the public, which began in the 1940s, have not taken all my time, as a list of my scientific publications demonstrates. More than two-thirds of them have been on subjects pertaining to struc-tural chemistry, crystallography, and biochemistry. Many of my papers on biomedical and epidemiological investigations are strictly scientific, so they may be too difficult for most nonscientists to comprehend readily.

It is the scientists who originate and develop knowledge of the physical world, enabling people to improve their lives through advancing techno-logies. However, material transformations initiated by some new con-cept or discovery may not necessarily benefit humans or the world around them; sometimes they end up harming them. Nor do we always use information that scientists have gathered through the centuries to improve our health or the way we live together. More than ever before, scientists should propose ideas for achieving the optimum health of that prime human invention—the social organism called society.

None of us knows how posterity will regard us once we are gone and our work is finished. We must leave it to others to evaluate the contributions we have made to knowledge of ourselves and our world. I hope you, the reader, will find these selected words of mine informative or provocative . . . and some of them, perhaps, even inspiring.

Deer Flat Ranch
Big Sur, California
May 1994

Preface

by BARBARA MARINACCI

Linus Pauling is the only person thus far to receive two unshared Nobel Prizes—for Chemistry (1954) and for Peace (1962). They reflect his dual commitment to society. Pauling lived and worked principally as a scientist, but he devoted considerable time and energy from the mid-1940s on—a half century in his long life—to discussing problems created by humans that may ultimately endanger the health or life of every person on earth. At the same time, he always proposed countermeasures for and solutions to these societal problems. He also suggested ways in which individuals could improve their own lives, including their health.

In his own lifework, Pauling integrated the sciences and, in the process, made them relevant to humanity. There is a remarkable flow from one line of scientific pursuit to another; likewise, his scientific interests and accomplishments led him to consider the relevant needs of and perils to his own species, humankind, that might be responded to by scientists—or people who at least can think like scientists. He believed that scientists, who are taught to approach problems objectively, should take an active part in advising and redirecting the society around them. He also believed that all people, no matter what work they did or would someday do, should be well educated in science. Not only would they learn to think clearly and purposefully, but they would also acquire a deep appreciation for all the wonders of earth—and therefore be unwilling to see it desecrated or destroyed.

Linus Pauling was fascinated by the unending marvels of nature, which included inanimate and animate, macro (the universe out there), micro (subatomic particles), and anything in between (crystals and humans). As he grew older, he was troubled that science got blamed for making people's lives more complex and dangerous because of its contributions to technology. He decided that the human capacity for rational problem solving was inadequately applied to designing and regulating an equitable world. With Pauling, justice, ethics, peace, equality, liberty, and integrity were not just principles; they were active watchwords he tried to live by and also wished to instill in others. In his view, science, which had already contributed greatly to human cultural evolution, now offered the means to remedy whatever was wrong in society. He sometimes declared that there was, or should be, a "science of morality."

Linus Pauling's research as a chemist began with determining the molecular structure of inorganic compounds and the nature of the chemical bond that holds atoms together in molecules and crystals. Once he had satisfied a desire to explain how these connections occur in inanimate matter, he applied his accumulated knowledge to organic substances. He then moved into the realm of biology, the basic life science that studies the structure and behavior of life in all its forms.

Pauling's background in chemistry and physics guided his investigations and discoveries. By applying his structural interests to the smallest divisions of life, especially to the complex identities and forms of protein molecules, he was instrumental in revolutionizing not only biochemistry and biology, but also medical science. Many scientists consider Pauling the founder of the new science of molecular biology, which furnishes the means for finally understanding how life came about and why it changes, and also how human health may be improved and extended through future research.

Pauling was a part-time mathematician, as most scientists must be. He loved doing calculations. (In later years a handheld electronic calculator replaced the ever-ready slide rule of the past, but he never really took to computers, though he recognized their usefulness and attraction to others.) Sometimes he quoted E. T. Bell's declaration that mathematics is the queen of the sciences but is also their handmaiden, which is true of the social sciences now as well.

Chemistry, physics, and biology—these three sciences, together with mathematics, underlay the world as Linus Pauling first explored it. His work in the scientific arena was conducted in the extensions and combinations of these four disciplines. Each of them in Pauling's grasp ultimately had applications, direct or indirect, to the human condition in his lifetime. Pauling made many discoveries. He began much scientific work that others would take up and bring to fruition. To the public nowadays he is perhaps best remembered for his interest in nutrition and human health, but at midcentury he was notable for his peace and anti-nuclear-testing activism—campaigns undertaken with the focus of the scientist and the zeal of a reformer. He provided continual observations about human society that merit attention from social scientists, economists, health practitioners, psychologists, educators, legislators, diplomats, and philosophers. His idealism would benefit politicians.

Despite his clear and unique genius, Pauling never claimed to possess an extraordinarily high IQ. He explained much of his success on the basis of his thinking more and working harder than most other scientists. Also, because of his educational background and research training, along with lifelong self-instruction in all sorts of fields that intrigued him, he had a comprehensive, interdisciplinary knowledge that often provided insights and syntheses unavailable to others. He could apply intense concentration—passion, really—on whatever he was doing or thinking. He perpetually enjoyed himself, partly because he never lost the child's pure joy when discovering something new about a beautiful object or a mysterious phenomenon.

Pauling excelled at both writing and talking. His total output of written and spoken words was—to use a much-favored adjective in his vocabulary—astonishing. His words have been preserved in various forms—handwritten, typed, carbon copied, photocopied, computer processed, printed, audiotaped, videotaped. Over one thousand printed documents have been counted, but Pauling also produced many manuscripts, complete or in fragments, yet to see identification and publication; thousands of letters, doubtless many not yet safely catalogued in repositories; interviews that may or may not have been transcribed; outlines, including commentaries on slides sequences, for lectures; random self-notes dictated or handwritten. Because of the importance of Pauling and his work, eventually all such materials should someday become available to scholars who wish to study particular aspects of this complex man—who despite his brilliance and constant accomplishment always was accessible. His winning simplicity, directness, and ease in demeanor are often on display in his writings and talks.

Needless to say, it has been both a challenge and daunting task to select from this vast wordage a collection of representative statements by Linus Pauling that will reveal his many facets as scientist, peace activist, health educator, and moral philosopher. Few details about Pauling's personal life have been provided here, although it was surely an important background to his work. This book is intended as an anthology, not a biography. (The Chronology in the appendix may help to convey main events in Pauling's life and work.) Excerpts from Pauling's writings and talks have been chosen to allow him to discuss different pursuits in science and society. Some inevitably reveal various roles he took (notably, researcher, educator, reformer). The four parts contain the major areas of Pauling's interests and concerns; chapter by chapter, the progression is rather chronological, showing the evolution of his work and thinking. However, there is inevitable overlap because during periods of his life several intense but widely disparate activities went on concurrently—protein studies and antinuclear protests, brain-function research and antiwar activities, peace quests and orthomolecular medicine. It was an extraordinary balancing act.

Pauling had a changeable voice when he wrote or talked; words, contents, and tone all depended on his audience, the subject undertaken, his mental preoccupations in that period, and even his mood. (Though an eternal optimist and even-tempered, sometimes he sounded gloomy indeed; he also occasionally became noticeably angry, especially over any kind of injustice.) He liked to reminisce about events in his life, so favored stories abound in the archives. Compared with most other people's, his memory was prodigious; yet the same tale might be told differently each time—emphasizing particular educational contexts or dramatic subtexts, giving variable details, and even yielding disparate messages. And being only human, when remembering something, he occasionally distorted, dramatized (he relished well-told tales), made small errors, or simply forgot, especially in old age.

Actually, Pauling rarely abandoned any keen interest; once taken up, it might lead him into a larger area that would then incorporate it, or else it would be resumed periodically, as time and circumstance allowed. Pauling's scientific pursuits, convictions, and commitments had a remarkable consistency and logical progression, all the way from structural chemistry to biochemistry to nuclear cautions and peace to orthomolecular medicine.

Pauling's first published article, on the manufacture of cement in Oregon, showed his respect for the precise word and fact, thorough research, and even some promotional enthusiasm. Indeed, had he not become a scientist, he might have made an excellent journalist or public-relations writer. (Ever intrigued with the evolution of both facts and language, he made a hobby of collecting old encyclopedias and dictionaries.) The serious young man of twenty-one who delivered the customary inspiring senior oration at the commencement ceremony in a land-grant college in Oregon—idealistic yet pragmatic, ambitious for professional achievement, altruistic about somehow benefiting society, eager for personal happiness—still dwelled in Linus Pauling at age ninety-three. He had been there throughout his life.

As both woman and editor, I feel that I must note and apologize in advance for one tendency in Pauling's word choices. In this postfeminism era it is strange indeed to see Pauling's near-invariable use of "he" and its variants when he is referring to a student or scientist; and in plural form, his "they" usually turn out to be "men"—whether chemists, politicians, or physicians—just as "man" is often used for singular "anyone" or for the human species itself. There is brevity and convenience in such words, but some readers may regard them as offensive or dangerous —vestigial and indelible proof of male territoriality. Yet these word artifacts usefully reflect history itself—certainly the entire history of science, in which women rarely participated. There were few women in science even when Pauling started out; Caltech, for instance, did not admit females as graduate students until the mid-1950s—their entry reportedly was thanks to Pauling. Women of course had long been on campus as secretaries, research assistants, and food preparers. The first female faculty member was hired in 1969, and in the following year female undergraduates arrived. No wonder, then, that before the mid–twentieth century few women pursued serious careers: Pauling's and other men's science articles and textbooks—and most teachers—were telling them, through seemingly inconsequential word choices, that they couldn't become professionals. Pauling greatly admired a number of female scientists of the past and among his contemporaries. Certainly he liked women, and he often acknowledged that his wife, Ava Helen, was the most powerful influence in his life. But though, apparently, finally alerted to a sexist bias in his words, he never appreciably changed his presentation style, which editors had not challenged earlier. The exclusivity was not deliberately insensitive; it was simply a product and indicator of Pauling's generation and all those preceding it.

Because this book avoids detailing work of Pauling's that is technical or abstruse, it cannot say to scientists much about the sciences themselves that Pauling delved into. However, some of them may be interested in reading his introductory statements to students, reminiscences, and lectures to laypersons. Also, they along with other readers may be interested in knowing more about how and why Pauling's dedication to science expanded to deep concern that the species that had created science thousands of years ago now seemed intent on self-destruction—maybe taking the rest of the planet with it, too.

My primary aim has been to display the range and main ingredients of Pauling's scientific delvings and humanitarian pursuits by providing his own words for what he thought, felt, did, or wanted to communicate to others—at a particular time in his life or in reflection afterward. Within him the researcher, the teacher, and the preacher were inseparable. All were inspired and sometimes even driven; each one activated the others. In some writings and talks only one may be viewed; in others, two or even all three. Depending on their own interests and perspectives, Pauling's listeners and readers, even his friends and associates, might perceive and respond variously to different Paulings—not necessarily valuing all of them equally.

Now, a personal note. I am not a scientist but have always been interested in science. I have scientists in my family, and at present I work happily among scientists. I first met Linus Pauling when I was sixteen years old, about to go away to Reed College, which his daughter too would attend. (Some years afterward she married my brother Barclay Kamb.) In 1958, when I was an associate editor at Dodd, Mead in New York, I persuaded him to write the book *No More War!* as the best way to alert the public to his concerns about nuclear warfare and nuclear testing.

Since 1992 I have been a consultant at the institute Pauling cofounded in 1973, which specializes in nutrient-oriented research and education. From the start, I was amazed by the many phone calls and letters that came into the Linus Pauling Institute, often addressed to Pauling himself. Some were desperate appeals for information that might improve some health problem. (His nature propelled him to try to help whenever he could, or to see that someone else would, especially if the condition was life threatening or baffling to physicians.) Many people, though, just wanted to thank Pauling and his institute for some previous assistance. Touching testimonials credited him with an improved chronic health problem, even a miraculous-appearing cure from a terminal illness. This grassroots evidence, coming in anecdotally from many hundreds of persons, cannot be faked. It shows that Pauling's orthomolecular approach worked well for some people, whereas conventional medicine's offerings seemingly had not. Pauling's renowned charisma probably provided some curative charms, true. But when children, even pets—unaware that some megavitamin or micronutrient regimen had been introduced into their diet—were benefited or healed, there could be no placebo effect. Having read by now many such letters, I have become something of a true believer in Pauling's health credo. But I am always aware (as a realistic counterbalance) that people are not apt to report failed attempts at gaining better health.

Increasingly I searched for the motivation behind Pauling's taking up the cause of orthomolecular medicine, and of vitamin C in particular, rather than pursuing pure theoretical research, as most scientists possibly expected of him in his declining years. I soon discovered the long history of his intense involvement with the chemistry of the human body —not just structural chemistry. Orthomolecular medicine was clearly the logical consequence of Pauling's biochemical and molecular work, combined with his inveterate promotional altruism.

As is often remarked, Pauling had an instinct for making discoveries in science. He also knew something about human health. Perhaps there was intuition involved, along with hyperbolic expression in getting points across—part of Pauling's style, anyway, no matter what he talked about. Largely, however, Pauling grounded his orthomolecular views and statements on recent research findings, some his own but mostly other people's, as well as on his previous work in and knowledge of biochemistry, physiology, and molecular structure and bonding.

Orthomolecular medicine applied Pauling's biochemical theories on physiology. In *in vivo* testing, people can choose to become subjects of their own experiments, in which before-and-after make the best test. (After all, for many centuries medicine had been a pragmatic art; what mattered was whether some method or drug worked effectively.) The many grateful letters that Pauling received and that his institute continues to receive attest to the validity of his approach to achieving optimal health and longevity.

Linus Pauling's death in August of 1994, midway in the construction of this book, was not unexpected. He had been in increasingly frail health for several years. After the news of his departure was swiftly circulated around the world, the institute was inundated by condolence messages that came by fax, telephone, mail, and personal visits. He will be missed by everyone who admired his contributions to science or his courage in taking up human causes (political, social, medical) that made him controversial—and often both.

As editor and writer on this project, my working metaphor has been to assemble a mosaic of the multifaceted life of Linus Pauling as it moved through time—with the story to be told as much as possible in his own words.

Linus Pauling Institute of Science and Medicine
440 Page Mill Road
Palo Alto, California 94306
April 1995

Editor's Note

The selections in this book, most of them excerpted, have come from more than one hundred separate sources, including publications and a number of unpublished manuscripts, notes, and interviews. (Approximately one-fifth of the Pauling selections are first being published here.) Since few of the texts used identical or even similar editorial styles, it seemed reasonable to establish one for this collection and attempt to maintain it as consistently as possible.

Because this volume is intended for general readership, not for scholars, certain minor editorial changes have been made in Linus Pauling's published (printed) writings and transcribed talks prepared by other editors. An attempt has been made, for example, to establish uniformity throughout in such minor matters as numerals, capitalization, hyphenations, spelling variations, and punctuation. In converting from publications printed originally in England, American spelling is reestablished (e.g., "color" instead of "colour"); Pauling's original manuscript, after all, would have used his American way of spelling.

Internal references in the Pauling texts to materials that no longer appear with them, such as drawings, figures, tables, and other chapters, have been eliminated. Subheads were excised. So were footnotes and endnotes. Readers who wish to obtain these references (which are surely important for research purposes) can consult the original version. In most cases, briefly cited papers and books within the text (author, date of publication) have been retained. The extracts and smaller quotations within the text are identified in the Notes on Sources.

Pauling (or an editor) sometimes used ellipses stylistically, to create a pause in his discourse. However, most ellipses in his texts here indicate that something has been removed—several words, a sentence, or many paragraphs.

Originally I chose a great number of potential selections from the huge collection of Pauling's written materials, and also some verbal ones. Gradually I narrowed them down, selecting representative statements about facets of his life, work, and thinking—whether made at the time or reflected upon later. The many inevitable redundancies had to be eliminated as much as possible. Maintaining some chronological and topical order also (regrettably) necessitated interrupting important lectures and articles. Then, to reduce the first-draft manuscript by two-thirds, so as to end up with a manageable size, we—Bob Bender, the editor at Simon & Schuster, and I—either removed entire pieces (and my explanatory text) or else excised parts—some of them large chunks—from the remaining ones.

In the process, not always has some choice article or passage been saved from the cutting-room floor. Nor will all aspects of Pauling's work be presented, let alone discussed. We hope, however, that enough remains of his own words to give readers a strong sense of who Linus Pauling was and what he did . . . and plenty of clues about what he thought and felt about the world we live in and our future in it.

I

The Path
of Learning

1901–1922

One

Starting Out

*Possibly I am a scientist because I was curious
when I was young.*

Linus Pauling always liked talking or writing about his childhood—
particularly about how it foreshadowed his career as a scientist. In various articles and in many interviews and public talks, he reflected on his
lifelong curiosity about the world, his interest in learning—and the joy
he took in discovering something he, and perhaps everyone else as well,
didn't yet know. He spoke often and to many people about these things
throughout his professional life, because he felt others—young persons
particularly—would learn from and even be inspired by his own experiences. Biographers, oral historians, and journalists planning profiles
often conducted in-depth sessions with Pauling. He was extraordinarily
open to such intrusions upon his time, particularly in his later life.

This account of Pauling's early years draws on various things he said
at different times, in about a dozen separate places, and blends short
passages selected from these sources into a single, uninterrupted narrative, told in his own voice.

Here, then, is an introduction to the young Linus Pauling.

I have always wanted to know as much as possible about the world. As
a child, I enjoyed reading and going to school. My early interest in
learning comes from an extreme curiosity about many things, which
remains my main driving force.

Possibly I am a scientist because I was curious when I was young. I
can remember being ten, eleven, twelve years old and asking, "Now why
is that? Why do I see such a peculiar phenomenon? I would like to
understand that." I suppose all children do this to some extent, but I

think that this feeling of curiosity was especially strong in me and probably is especially strong in most scientists.

Scientists are fortunate people in that there's so much more in the world that they can appreciate and enjoy than what people who don't have some understanding of science can appreciate and enjoy. So the world is a richer place for scientists than for nonscientists, and I feel sorry for all the nonscientists.

Usually I say I am a chemist; sometimes I say I am a physical chemist. My Ph.D. degree, from the California Institute of Technology in 1925, was given to me for my work in physical chemistry and mathematical physics. Perhaps I should characterize myself as a physicist with an interest in chemistry. My scientific work, however, has not been restricted to chemistry and physics, but has extended over X-ray crystallography, mineralogy, biochemistry, nuclear science, genetics, and molecular biology; also nutrition and various aspects of research in medicine, such as serology, immunology, and psychiatry.

So, more recently I call myself simply a scientist. I am also inclined to be a theoretical scientist, because I work with ideas—with speculations based on already observed phenomena or with concepts that can be put into quantifiable and mathematical forms. In later years I have done this kind of work much more than dealing directly with the actual matter, particles of energy, or biochemical processes that my speculations and findings pertain to.

I was born on February 28, 1901, in Portland, Oregon. My parents, Herman Pauling and Belle Darling, had married the year before. My great-grandparents and grandparents were pioneers who had traversed the Oregon Trail, to settle in Oregon Territory. In 1905 my parents went to live in Condon, Oregon, on the eastern side of the Cascade Mountains, where my mother had grown up. The population of five hundred included mainly cowboys, ranch hands, and saloon keepers. I started school there, but in 1910 our family moved back to Portland.

When and why did I decide to study science? I really can't say exactly. I don't remember that I ever wanted to be a great baseball player or locomotive engineer. My father was a druggist, and as a young boy I used to hang around his store. I remember watching him mix up different pharmaceutical concoctions for his customers and then put them into little bottles. Perhaps at that time I wanted to become a druggist myself; I can't remember what I thought. My father died when I was nine years old, so he couldn't have had a very great effect on me. He died rather suddenly of peritonitis, which was a great shock to my mother, who now had to find ways to support herself and her three children. I had two younger sisters, Pauline and Lucile. As I grew older, my mother began to depend on me to contribute to the family income.

There is some evidence, though, that when he was alive my father had been pretty interested in me. I don't remember much about it, but I do know that he was impressed with my early reading ability and intellectual curiosity. Some months before his death he wrote a letter to the Portland *Oregonian*, asking for advice as to what books he should get for his precocious young son, who was an avid reader. "Please don't suggest the Bible and Darwin's *The Origin of Species*," he said, "because he's already read these books." He described me as being especially attracted to ancient history and natural science.

I read a great many books when I was a boy. I occasionally went to visit the family of an uncle who was a judge. While with them, I spent much of my time reading the *Encyclopaedia Britannica*. Finding it wonderful, I sometimes read sections aloud to my girl cousin, who was younger than I. I also remember reading books on ancient history and *Alice in Wonderland* and *Through the Looking Glass*. Examining a Doré-illustrated edition of Dante's *Inferno* made me skeptical about revealed religion. My mother was a Congregationalist. I was baptized when I was three years old and often was taken to church. But by the age of twelve I decided that I could not believe many of the stories of miracles in the Bible. I kept this opinion mostly to myself for a long while.

Even as a child, I wanted to understand the world about me. I still recall being a small boy in Portland and walking along in the rain. When I looked up through my umbrella at an arc light about a block away, I saw a white spot at the center, where the arc light itself shone through; and also some colors—a sequence of violet, indigo, blue, green, yellow, orange, red, with the array of colors in a rainbow, which I knew was called a spectrum. There were eight of these spectra: four rather close in to the white spot, one to the right, one to the left, one above, and one below, and four others somewhat farther out, in between these directions. I puzzled over this observation.

I had no idea about the cause of this intriguing optical phenomenon, but I remember thinking that I would probably learn the explanation later on, in the course of my studying various subjects in school. And I did. The pattern of colors was caused by the interference and reinforcement of the light waves which passed through the meshes of the fabric, and the angular spread of the pattern was determined by the ratio of the wavelength of the light to the distance between the meshes of the cloth.

Knowing the wavelength of the light, one could use the observed pattern to calculate the distance from one thread to the next. One could also infer from the nature of the diffraction pattern the type of weave, and thus determine the structure of the cloth.

Long afterward, I would find out that the investigator of molecular structure follows a similar procedure. He may, for example, use a beam of electrons—instead of the rays from the distant arc light. The molecules of the chemical compound which he seeks to measure are analogous to the fabric of the umbrella. When the beam of electrons passes through the molecules, the electron waves are scattered into a diffraction pattern which is photographed—and by analyzing this photographed pattern the scientist learns how the atoms are arranged in the molecule and how they are spaced with reference to one another. This is known as X-ray or electron diffraction.

Is scientific talent something people are simply born with? I suppose so: the geneticists and psychologists have studied this matter, the inheritance of musical ability or mathematical ability and so on. Most of them seem to feel that your genetic makeup determines to some extent what your interests and abilities will be. But it's quite clear that environment, circumstances, also have an effect. If a student is doing mathematics and doesn't understand something, he might be sunk from then on. Mathematics is built up as a logical structure, and if you come to a point where you don't understand, from then on you may say, "Well, you see, I wasn't meant to be a mathematician." But if a person who doesn't understand something manages to overcome that and then to understand it and go on, pretty soon he may, like Einstein, feel that there's no problem that is too difficult for him to decide not to try to solve.

I've said that the passing grade in mathematics should be 100 percent. Suppose it's 70 percent; then you go on to the next year, and again you only understand 70 percent of the 70 percent that you did understand, or 49 percent. The following year, you understand 70 percent of that, or 35 percent, and so on. So the student gives up and says, "I'm just incapable of understanding!"

I was fortunate that this didn't happen to me. In fact, I had the feeling that I could understand everything! And I continue to be astonished at the discoveries that scientists and others continue to make about the nature of the world.

I've retained this feeling that if I try hard enough, I can understand something. When I read *Nature*, here's somebody reporting something, and if it fits into my picture of the universe, OK, I won't think anymore about it. But if it doesn't fit in, then I ask, "Can I find out something new by seeing why it doesn't fit in?"

When I was a boy, I began formulating a picture of the universe, and there were parts of the universe I didn't understand and didn't make any effort to understand. But there were also parts of the universe that I tried to understand.

I think I was born with an exceptional memory, and I have acquired a tremendous background knowledge of chemistry and related subjects. I was fortunate in having good teachers who didn't put me off learning, so that my curiosity was not quenched, and I was not given the impression that there were any limits to my understanding. So there's an element of chance involved here: if you have a good teacher, you may get by that particular hurdle. If you have a poor teacher, just a matter of chance, then the nature of your life from then on will be changed by the fact that here was something that you weren't able to understand.

Still, young people can also pursue their own special interests. They can start learning things on their own, when challenged by something in their surroundings; also they can get inspired from reading. They can create their own laboratories and field studies, conduct experiments to test what they would like to find out—as I did. Later on, if their enthusiasm is maintained along with information they have gathered and questions not yet answered, they will probably find the knowledgeable mentors they need, if they are in some kind of educational setting.

When I was eleven, with no outside inspiration—just library books— I started collecting insects. Not only did I collect insects, I also read about insects. I was filling my mind with a lot of information about the Lepidoptera and Diptera and so on. At the time, I was interested only in insects! Which is why, before I got interested in chemistry itself, I began to need chemicals.

"A person who collects insects needs to have a killing bottle," I said. I got a Mason jar from my mother. All I needed now were 10 grams of potassium cyanide and perhaps 50 grams of plaster of Paris. William Ziegler, a druggist who had been a friend of my father's, gave me what I needed. I took the chemicals home, went out on the back porch— because I knew that potassium cyanide was dangerous—and dumped the poison into the bottle. I mixed the plaster of Paris with some water, put it in the bottle on top of the potassium cyanide, and let it harden. I now had my killing bottle. I collected a lot of insects afterwards.

Just think of the differences today. A young person gets interested in chemistry and is given a chemical set. But it doesn't contain potassium cyanide. It doesn't even contain copper sulfate or anything else interesting because all the interesting chemicals are considered dangerous substances. Therefore, these budding young chemists don't have a chance to do anything engrossing with their chemistry sets. As I look back, I think it is pretty remarkable that Mr. Ziegler, this friend of the family, would have so easily turned over one-third of an ounce of potassium cyanide to me, an eleven-year-old boy.

For some reason, after a while entomology, the science of insects, did not satisfy me intellectually. When I was twelve, I began reading about rocks and minerals and started collecting them. I occasionally walked by a big house that was less than a mile from my home. In the center of an acre or two of land, it was surrounded by a newly built rock wall. The rocks were granite and there were pieces of it, two or three inches in diameter, lying on the ground after the stonemason who built the wall had chipped them from the granite blocks. I picked up some of these granite chips and looked at them carefully. In the rock there were three kinds of crystal grains: white or transparent grains, pink grains, and black grains, the black grains being flat plates. These grains were for the most part around an eighth or a quarter inch in diameter. From my reading I knew that the white or transparent grains were quartz, the pink grains were a form of feldspar, and the black grains were mica. I was actually seeing tiny, somewhat crude crystals.

What is a crystal? It is a form of matter in which the component atoms are arranged in a regular, repeating way like bricks in a brick wall. Because of this arrangement, a crystal that grows without interference from surrounding objects is bounded by flat faces of definite shape— triangular, rectangular, quadrilateral, etc.—at definite, fixed angles to one another. A crystal has different properties in different directions— for example, the property of cleavage, which is a tendency to break along a definite direction, parallel to a certain crystal face, regardless of the size of the crystal fragment. An excellent example is mica, which can be cleaved into thin sheets parallel to a single planar direction in the crystal. Mica crystals in granite have the shape of roughly hexagonal tablets, in which the direction of cleavage is parallel to the flat plane of the tablet.

As for mica, I was already familiar with it in a different way, from a firsthand impression I got when I was about five years old. I had been looking out the window in Condon, Oregon, at the snow on the ground. My mother was trying to get me to get dressed, and I had stopped— perhaps taken off my pajamas—but I had started looking out the window. She said, "I've told you several times to get dressed. Put your clothes on now—go on, do it!" and I don't think that she really swung at my behind, but I thought she was going to do it, so I ran, and ran into a hot stove which was in the room, a wood stove to give us heat, and got a burn on my tummy. Perhaps I touched the mica of the stove window. In those days, cleaved sheets of mica were commonly used for stove windows instead of glass.

I was curious about granite and other rocks, but it seemed to me that before I could understand rocks I needed to understand minerals. Surely if I had an understanding of the nature of quartz, feldspar, and mica, their structure and their properties, that understanding would be basic to an understanding of the nature of rocks. Accordingly, for a year I read about minerals and attempted to develop some understanding of their nature. Now minerals began to interest me more than rocks.

One thing that I learned from reading books about mineralogy was that minerals often have a definite chemical composition. For example, the mineral quartz was described as silicon dioxide, with one atom of silicon combined with two atoms of oxygen. Obviously, I would need to find out more about elements and compounds—what they were basically, where they were found, and how they fit together.

I had entered Washington High School about the time I started collecting rocks and minerals. I often walked home from school with my friend Lloyd Alexander Jeffress, who was my age. One day, when I was thirteen years old and in my second year of high school, Lloyd asked if I would like to come up to his room and see some chemical experiments. So we went to the second floor of his house, and he carried out several chemical demonstrations that I found intensely interesting. The most impressive one was when he mixed sugar and potassium chlorate in a ceramic bowl and then poured sulfuric acid over the white powder, immediately inducing a chemical reaction that produced a lot of steam and a pile of black carbon.

Until then I hadn't thought about the fact that substances can be converted into other substances. For instance, I just accepted combustion—the burning of wood in the stove—as part of the everyday world. I hadn't thought about the general phenomenon of the conversion of one thing into another. "I am going to be a chemist!" I decided right at that time.

So when I reached home, I found my father's old chemistry book and began reading it. I also carried out a manipulation—consisting, I think, only of boiling some water over an alcohol lamp. But it made me feel as if I were already on my way to becoming a chemist. Later, I pored over various books and began to learn what was generally known by the early years of the twentieth century within the science of chemistry. I started reading about its history, and then tried to understand what the chemists' ideas and problems and discoveries had been at the time they lived.

I began carrying out my own experiments at home, at first using that old chemistry book that had belonged to my father. I set up a laboratory for myself in the basement and started getting my own equipment, such as flasks and beakers. Mr. Ziegler, the druggist who had been my father's friend, now gave me chemicals and some apparatus. Also, a man who lived next door to us, who worked as the stockroom keeper at a dental college, brought home many pieces of glassware for me; they were all chipped and would otherwise have been discarded.

There was another way I got chemicals. Almost every weekend I took the train to Oswego, seven miles from Portland, to stay with my grandparents. My grandfather was a night watchman at the foundry there. I would often go down there with him, and then go off to an abandoned smelter, about a quarter mile away. It was a wooden structure that was falling down; the laboratory roof had collapsed. But there were hundreds of bottles of chemicals and ore samples. For example, I got a two-and-a-half-liter bottle of sulfuric acid and similar bottles of nitric acid and hydrochloric acid. The sulfuric acid was black; I suppose the bottle had been opened and a little organic material had fallen in and been dehydrated. There was also a big bottle of potassium permanganate. I'd take these things back with me on the train to Portland, then get on the streetcar to travel the two miles to my home. It was really great to get those chemicals!

I had made a rough wooden workbench, and had bottles of chemicals sitting at the back of the table, including a Winchester bottle of concentrated sulfuric acid. One day I started to do an experiment requiring this acid. As I replaced the bottle, I banged it against the concrete wall and it broke, releasing about 5 liters of the acid. After I had taken off my clothes and washed myself off, I got a broom and swept the acid over to the drain. It seemed not to react with the concrete floor, but when I hosed it down with water it effervesced vigorously. Only much later did I learn about the ionic theory and the hydronium ion. My earlier experience strengthened my appreciation of the theory.

I still saw a good bit of Lloyd Jeffress, who was my best friend. But we didn't do any more chemical experiments together. He ended up being a professor of psychology at the University of Texas and remained a good friend throughout his life. Well, he was a psychologist all right! When we were about fifteen, we were together at my grandmother's house and she asked, "Liney, what are you going to be when you grow up?" I said, "I am going to be a chemical engineer." Then Lloyd said, "No, he isn't. He is going to be a university professor."

I had no expectation then of becoming a professor someday. I simply thought I would be a chemical engineer. I didn't have much knowledge about professions. But I did know about engineering and chemical engineering because a cousin of mine was studying highway engineering at Oregon Agricultural College, and I knew that they taught chemical engineering there. Lloyd Jeffress would be going to OAC too, and he and his aunt and uncle, who were his guardians, encouraged me to go to college. In fact, they all insisted on it.

OAC had originated as one of the land-grant colleges, so it prepared people for practical, technical occupations like chemical engineering, which I unquestioningly believed was the right profession for me. Anyway, in those years OAC—which became Oregon State College and finally Oregon State University—didn't yet offer a major in chemistry.

I also decided to go to OAC primarily because it was the only college I could afford, since we didn't have any money and the tuition was free. But I'd have to work on the side, of course, to pay for my books and living expenses. Nobody ever suggested to me that I apply for a scholarship somewhere. My high school grades were not outstanding except in the classes that really interested me, such as mathematics and science. The family financial situation was difficult, so my mother certainly would be unable to finance my college education. Still, I wanted to go on with my studies. And because I had done well in school, my friends urged me to continue.

During my primary and secondary school days, many teachers had made significant impressions on me. In the main, they were teachers of mathematics, physics, and chemistry. I remember especially vividly Mr. William Greene, the teacher at Washington High School who taught the chemistry class, which I took when I was a junior. I think he had a master's degree from Harvard. He was a very good teacher. Frequently, toward the end of the school day, he would ask me to remain an extra hour to help him in determining the calorific values of the coal and oil the municipal school board purchased. It was probably largely a way of giving me a little extra instruction.

The year after I had taken his course in chemistry, Mr. Greene allowed me to work in the chemical laboratory by myself. I used several textbooks, but he would sometimes give me special problems to do in analytical chemistry, qualitative analysis. One day he gave me an unknown substance and said that I should remember the words *hoi holoi strategoi* and tell him what they meant when I reported the results of my analysis to him. I hadn't started studying Greek yet; I got White's *Elements of Greek* and learned some Greek from it. I learned that his words meant "all the soldiers." He had put all of the metals into the sample he gave me to analyze. When I got my high school transcript I saw, to my surprise, that he had credited me with an extra year of high school chemistry.

I was eager to finish up high school and start college, to prepare myself for a career. To complete the requirements for graduation I needed to take a history course for a year, but in the spring of my senior year they wouldn't let me take the two semesters of the class concurrently. So because I lacked one semester of history, I didn't graduate before going off to OAC. They had no requirement for a high school degree so long as one had enough course credits. (Years later, after I had made a name for myself in science, Washington High School in Portland awarded me an honorary high school diploma.)

I had always had small jobs to earn money that helped my mother out at home. Now I needed to earn more money and save some for college in the fall. That summer I took a job with a local machine shop. My employers were very pleased with me, as a responsible young person who could learn fast. Every week they increased my paycheck; then they offered me a regular position at $150 a month. My mother, a young widow with three children, of course wanted me to take the steady job —that was a good salary at the time—and forget my plan to go away to college. But I was determined, so at first she got pretty upset with me. I don't think she ever understood me very well. At any rate, she certainly didn't know anything about science or academic careers. Eventually she accepted my decision, even though the next few years were difficult ones financially for her as well as for me.

It may be that this early period of my life developed in me a sort of independence of thought and decision. I was never pugnacious. I was pretty retiring. It wasn't long, especially when I was in college, that I began to develop self-confidence, confidence in my own ideas. It showed up especially when I was a graduate student.

I was sixteen years old when I went away to Oregon Agricultural College in Corvallis. It was September of 1917—the year the United States got into World War I, which had begun in Europe three years earlier. Some of the young men who otherwise might have been in college had enlisted, and more would be drafted. Feeling patriotic and wanting to be prepared to fight right away if sent off to war, I signed up for the Student Officers Training Corps—what is ROTC now. I participated in its various training exercises while at OAC, and eventually I achieved the rank of major. I also went out for track as my main athletic activity; I never became a star athlete but of course I envied those who did!

At Corvallis I found ways to sustain myself while living in rented rooms off campus. When I was a freshman I worked at odd jobs in the kitchen of the girls' dormitory, such as chopping wood for the wood stoves and cutting the sides of beef into steaks and roasts. There was a period when I didn't have money enough to pay for board and room in the boarding house, but I paid for the room and then spent what money I had getting food. I didn't have enough knowledge of how to get along, how to buy bread and cheese, for example. The next year, the people in the chemical engineering department gave me a job as a stockroom assistant, preparing chemicals for the student laboratories and so on, and from then on I at least got enough money to live on and to give some to my mother.

It is difficult to evaluate this aspect of my college experience. Routine outside work took time away from my studies. And this may have been detrimental. Until I was given the part-time job at the college, I spent a hundred hours a month working elsewhere. This meant three hours a day chopping wood or performing some other kind of labor. Still, because it accustomed me to long hours of hard work, this experience may actually have been profitable.

I knew that I liked mathematics, and in my freshman year took virtually all the courses that OAC offered in that field, which were only elementary mathematics and calculus. It did not occur to me at the time that I might have continued learning just by studying mathematics on my own. I thought the only way to learn something was to have some teacher teach you in class. (Five years later I would quickly make up for deficiencies in mathematics, and physics too, when I started graduate school.)

The freshman chemistry textbook was Alexander Smith's *General Chemistry*. In it I first encountered the word "stochastic"—a method of scientific investigation and discovery that I would often later apply to my own methodology. It comes from the Greek word *stochastikos*, which means "apt to divine the truth by conjecture." As I remember it, Smith said that you make a stochastic hypothesis—for example, that a substance is a hexahydrate—and then you can immediately test the hypothesis by carrying out an analysis. In this case, you would have to remove the water to find out how much water was present.

During the summertime, between the school years, I worked at various jobs. In the first summer my cousin Mervin Stephenson and I worked at a shipyard near Tillamook. The next summer I worked for Oregon's highway department as a paving engineer, mostly on the eastern side of the state. I sent home most of the money I made to my mother, who was supposed to set aside some of the funds for my schooling. She was having such a tough time of it, though, that in the summer of 1919, after my sophomore year, I saw that I would have to compromise. I agreed not to return for my junior year at college but to continue working at my job with the highway department for at least a year before going back.

Somehow the professors in the chemical engineering department heard about my predicament. In the autumn they sent me a telegram offering me a full-time position teaching a sophomore class in quantitative chemical analysis. The war was over by then and the university's science classes were being filled by young men released from military service. Though I was still an undergraduate and no older than the students and often even younger than they, I had taken the class in the previous year and excelled at it. My work in the stockroom, mixing up chemicals for student analysis, had also put me in good stead for this job. I got the nickname of Boy Professor.

I remember puzzling over a difficult calculus problem as I walked about the campus in my [first] year and finally seeing how the problem could be solved. I had another lesson, of another sort, during my sophomore year when employed by the chemical engineering department to make solutions of different chemicals for use in the laboratories. When we needed some concentrated ammonium hydroxide solution, I bubbled ammonia gas through distilled water in a large carboy. I had the problem of transferring the saturated solution to smaller bottles. The carboy was too heavy for me to lift, so I made a siphon by putting two glass tubes in a two-hole stopper in the carboy, one of them to allow the flow of the liquid and the other to permit me to blow into the carboy to develop enough pressure to start the siphon. After the siphon had started I opened my mouth, forgetting that the ammonia gas above the liquid in the carboy was under pressure and would blow back into my mouth. The ammonia caused some of the mucous membrane of my mouth to fall off. This experience taught me that one must not be satisfied with having solved one problem but must consider the possibility that the solution of the problem will lead to other problems.

My desk as an instructor was located in the chemistry library, so I had easy access to all the new chemistry journals that came in, as well as back issues. That's how I first developed my interest in the nature of the chemical bond, in 1919. I read the *Journal of the American Chemical Society*. That year there were several papers by Irving Langmuir on the electronic theory of chemical bonding, based upon a paper published in 1916 by Gilbert Newton Lewis, the great chemist who was dean of the College of Chemistry in the University of California at Berkeley.

These two chemists were saying that the chemical bond consists of two electrons that are held jointly by the two atoms that the bond connects. I knew by then a great amount of descriptive chemistry, and I could see how the shared pair of electrons could explain what the forces are that hold atoms together. I also saw that the first steps were now being taken toward a chemistry that would be more than a collection of facts held together by a few empirical theories—that is, a real, systematic science of structural chemistry might be developed. Enthusiastic about these new ideas of Lewis's and Langmuir's, I gave a seminar on the nature of the chemical bond—my first scientific seminar.

A great deal of theoretical physics work, along with experimental work to test it, had been going on in research institutions for a quarter of a century now. Five years before I was born, the existence of the electron was discovered—the negatively charged particle that was making so many modern inventions possible, particularly those using electricity. So I came close to being born in that period of time before it was known that there are electrons in the world!

J. J. Thomson at Cambridge University discovered the electron; another scientist in England, G. J. Stoney, previously predicted that it existed as a discrete form of matter and actually suggested the name. In this marvelous period of only a few years beginning in 1895, modern physics came into existence. Actually, the nucleus of the atom itself, consisting of the positively charged protons (the number of which balance the number of electrons) plus uncharged neutrons, was not discovered until 1911, when I was ten years old.

In my last three years at OAC, I continued to spend much time reading the chemical literature, focusing with particular interest on the new structural chemistry. Yet in my early years of teaching elementary chemistry to college students, I would explain the chemical bond formed from the interaction of atoms in the standard graphic way that chemists, not physicists, often described it for simplicity's sake. I used to tell them that the chemical bond could be thought of as a "hook and eye." I would say, for instance, that the alkali atoms (sodium, potassium) have an eye, whereas the halogen elements (chlorine, fluorine) have a hook. The chemical bond consisted of the hook hooking into the eye. Why did I say that it was the halogens that had the hook? Because chlorine gas is Cl_2 and fluorine gas is F_2. At the time, it was thought that sodium vapor was just Na (a single sodium atom, no Na_2's). Whereas two hooks together, I said, could also form a bond. So you got F_2 and Cl_2. It really was a primitive picture of chemical bonding, but that was the state of the art—the understanding we had in those years of chemical bonds.

Not only was I interested in the nature of the chemical bond, but I was curious also about variations in physical properties of different chemical substances—properties such as diamagnetism and paramagnetism (responses to a magnetic field). I wondered about the molecular basis for electromagnetic phenomena such as the dielectric constant, in which molecules tend to orient themselves with positively charged ends pointed toward a negative plate, and vice versa.

For years I had mulled over the tables of properties of substances, magnetic susceptibility, hardness, and of course in the case of minerals, cleavage, color, and other properties. Magnetism was a property that especially interested me. I was also very much interested in metals and alloys, so when I was a senior at OAC I had the idea of carrying out an experiment to check on a hypothesis that I had formed. I knew that one could, by electrolysis, convert iron compounds into iron crystals, and the idea I had was that if I would deposit crystals of iron electrolytically in a magnetic field, the interaction of the magnetic field with the crystals should orient the crystals. Though I didn't know enough about this whole field of research to formulate a good experiment, I did attempt to carry it out. I wasn't successful. There were practical problems about the electrolysis that caused difficulty for me. (Interestingly, my first work in biochemistry, beginning in 1934, would involve studying the magnetic properties of iron in hemoglobin. By then, of course, I had a far better understanding of how to conduct experiments.)

I also wished to better understand the cause of such identifying features as hardness, cleavage, and color in crystals and minerals generally. Additionally, I strove to explain the structures of various crystals—like the quartz, feldspar, and mica minerals in the granite specimens I had collected as a boy; and crystals of other inorganic substances, too. I knew that some elements, such as carbon and copper, and many compounds, such as ice (frozen water) and salt or sodium chloride, in their purest forms exist as crystals. Looking at them, I might be seeing a vastly enlarged and opaque structure of a giant molecule. Perhaps I was already attracted to the realm of my future work.

For by then I had a strong desire to understand the physical and chemical properties of substances in relation to the structure of the atoms and molecules of which they are composed. This desire largely determined the course of my work for the next fifty years and more.

As I try to remember the state of my development at that time, I am led to believe that this desire was the result of pure intellectual curiosity and did not have any theological or philosophical basis. I was skeptical of dogmatic religion and had passed the period when it was a cause of worry; and my understanding of the experiential world was so fragmen-

tary as to be unsatisfactory as the basis for the development of a philosophical system. I was simply entranced by chemical phenomena, by
the reactions in which substances disappear and other substances, often
with strikingly different properties, appear; and I hoped to learn more
and more about this aspect of the world.

It has turned out, in fact, that I have worked on this problem year
after year, throughout my life; but I have worked also on other problems,
some closely related, such as the structure of atomic nuclei and the
molecular basis of disease, and others less closely, such as the pollution
of the earth with radioactive fallout from the reckless testing of nuclear
weapons, the waste of life and the earth's resources through war and
militarism, and the maldistribution of wealth.

During my delayed junior year, which began in the fall of 1920, an
engineering professor, Sam Graf, gave me a job correcting papers in the
courses he taught, about statics and dynamics, bridge structure, strength
of materials, and so on. I also helped him in the laboratory. I had taken
a class in metallography from him, which I greatly enjoyed. It started
me thinking intensively about the structures and behaviors of the metals, which I have often considered ever since.

A metal, as I had learned, is defined as a substance that has large
conductivity of electricity and of heat, has a characteristic luster, called
metallic luster, and can be hammered into sheets (is malleable) and
drawn into wire (is ductile); in addition, the electric conductivity increases with decrease in temperature.

Actually, about eighty of the more than one hundred elementary
substances found on earth are metals. The metals themselves and their
alloys are of great usefulness to man, because of the properties characteristic of metals. The development of civilization had almost as much
to do with mankind's learning about metals—about how to make tools
and machinery (and, unfortunately, weapons too)—as it did with agriculture, with people's learning how to grow crops. The craft of ex-
tracting pure metals from ores and then of combining them into alloys,
to be formed into useful shapes by casting and forging, was continually
improved. This really was early industrial chemistry, chemical engineering.

In the Middle East, in the fourth millennium B.C., alloys began to be
made using copper. Bronze, an alloy of copper and tin, preceded the use
of iron, which is why the first period in the development of metallurgical
technology is called the Bronze Age. Our modern civilization, the Industrial Age, is based upon iron and steel, and valuable alloy steels are
made that involve the incorporation with iron of vanadium, chromium,

manganese, cobalt, nickel, molybdenum, tungsten, and other metals. The importance of these alloys is due primarily to their hardness and strength. These properties are a consequence of the presence in the metals of very strong bonds between the atoms. For this reason it is of especial interest to understand the nature of the forces that hold the metal atoms together in metals and alloys.

By my junior year I was already thinking about going to graduate school. Chemical engineering no longer seemed very appealing to me as a profession. I was attracted more to ideas, to theories—to discovering and proving things, rather than applying them out in the "real world." During the course of educating students for four years, the OAC program exposed them to all sorts of useful trades and crafts. For two years, as an example, I took courses in the School of Mines. I also learned the principles of blacksmithing, so I would be able to do that, including affixing iron shoes onto horses' hooves, if no other employment opportunities appeared when I graduated. My summertime work with the highway department also proved I could do practical work when and if I had to. But my strong interests lay elsewhere, in scientific pursuits that required still more years of education. As before, my mother did not appreciate my desire to continue on with my schooling and not get a well-paid regular job in some industry, as she expected me to do after five years at OAC.

I had encouragement from others to continue my education, however. A professor of chemical engineering, Floyd E. Rowland, was a real enthusiast about graduate work. He was one of the few people on the OAC faculty who had a doctoral degree. There were twelve graduates in our chemical engineering class of 1922. Largely because of Rowland's influence, seven of us chemistry students from that small "cow college" went on to graduate work, and I think six later got Ph.D.'s. At that time, the chemical engineers were the smartest students at Oregon Agricultural College.

Now that I had decided to attend graduate school, where should I go? When I was a sophomore, the department head, John Fulton, had shown me an announcement from a science institute offering fellowships in chemistry to graduate students; he suggested that I ought to go there someday. I then read up on this school in southern California. Founded in 1891 in Pasadena, it was known as Throop Institute or College until 1920, when it name was changed to the California Institute of Technology. I had even thought of transferring there in my junior year. But when I wrote to Arthur Amos Noyes, the head of their chemistry division, about this, I said I would have to earn my own living. I then got a letter back from someone else, who told me it was impossible to attend there and earn a living at the same time, since the academic load was so demanding.

Still, I had kept CIT in mind as a good place for me to attend eventually. (That original abbreviation is now, of course, pretty much outmoded, and the shortened name "Caltech" is familiar to most people. However, I myself usually use "CIT" because I still much prefer it.) But there were other attractive possibilities at universities with innovative chemistry departments and excellent professors, such as Harvard, University of Illinois, and UC Berkeley. I favored the latter because G. N. Lewis was in charge of chemistry there. I knew I was already interested in structural chemistry, and he was an acknowledged expert.

What happened then was that first I was accepted at Harvard. They offered me a half-time instructorship but said it would take six years to get a Ph.D., so I turned that offer down. Then I heard from A. A. Noyes at CIT—who said I must reply with an acceptance of a fellowship immediately if I wished to go there. I hadn't heard yet from Berkeley, so I decided to enroll in the CIT graduate program, which would give me a doctorate in three years while employing me as a part-time instructor and researcher. Much later I heard a story—probably it's apocryphal— that when Lewis had looked over the several dozen graduate applications to the Berkeley chemistry department in early 1922, he came to one, looked at it, and said, "Linus Pauling, Oregon Agricultural College. I have never heard of that place." So my application went into the discard pile.

In my senior year at OAC a second opportunity to teach had been given to me. It turned out to affect the rest of my life! At the start of the Christmas period, while I was waiting at the station to board the train home, one of the professors came up to me and asked whether I would be interested in taking over as instructor of an introductory chemistry class for young freshmen women who were majoring in home economics—food science and nutrition. I agreed to do this.

Teaching that freshman course was how I met my wife-to-be, Ava Helen Miller. At the first meeting, I picked her name at random from the class list and asked her to explain the chemical action of the ingredients in baking powder. She answered my question quickly and correctly; she subsequently proved herself to be both forthright and very bright. In fact, I always considered her quicker and in some ways more intelligent than I—as a test we both took, early in our marriage, proved her to be. Not only was she quicker, but she had more correct answers. By then, of course, it was too late for me to seek a mate over whom I might feel mentally superior, as some men seem to want to do!

Sometimes the "chemistry"—as an intuitive attraction between two people is often called—is just right. Certainly this was so with us. Or at least initially for me, for it took a while to persuade Ava Helen to go out with me. This was partly because she was angered by my writing her a note saying that another teacher had been accused of paying too much attention to a female student. And later she resented my giving her lower marks than she really deserved; I thought this would demonstrate that I wasn't showing any favoritism among students. But once we started seeing each other outside of class, we soon became engaged. We wanted to get married that summer, before I would leave for graduate school, but neither of our mothers were in favor of this; they considered us too young. So after a summer of working again in supervising highway paving, I went off to California without Ava Helen, who entered her second year at OAC. We had vowed to write each other every night.

A couple of the OAC professors even lent money to help me travel to Pasadena and get set up, and I repaid the loans as soon as possible. I would be rooming at the house that the mother of a good friend at OAC, Paul Emmett (who one day would marry my sister Pauline), was going to rent in Pasadena. She wanted to watch over her son in his first year as a graduate student at Caltech, and therefore could extend her housemotherly attentions to me as well.

Throughout my childhood I expected to learn more about nature and how various natural phenomena are explained by continuing my studies and finding out what understanding human beings had amassed to explain their observations of the universe. Only when I began studying chemical engineering at Oregon Agricultural College did I realize that I myself might discover something new about the nature of the world, and also have some new ideas that contributed to a better understanding of the universe.

This thought would recur, and burst upon me explosively, when I became a graduate student at the California Institute of Technology in the fall of 1922. I soon recognized that the scientists there were working very vigorously on research problems to get additional information about the nature of the universe, and this really appealed to me very strongly.

Two

What Is Chemistry?

Chemistry is wonderful! I feel sorry for people who
don't know anything about chemistry. They are
missing an important source of happiness.

Linus Pauling had a lifelong passion for chemistry. There is no better
word to describe his devotion to the science he had chosen to pursue at
the young age of thirteen. He was always interested in discoveries in
chemistry, whether contemporary or historical. He often talked about
chemists of the past, early experiments, and milestone discoveries. Thus
he not only conveyed the drama involved—he greatly enjoyed storytell-
ing—but also showed how determined individuals contributed to scientific progress. He did this himself, and so might his listeners in their own
chosen area.

Pauling believed, of course, that anyone taking a chemistry course
would derive important benefits. At the start of his first textbook he
mentioned several, then explained why students would sometimes also
hear from him about the history of chemistry as well. (When the paper-
back manual was printed in 1941, Caltech was still strictly a male bas-
tion: hence the "he" was unavoidable.)

THERE are two main reasons for a student to study chemistry: first, to
learn what the science is, and what its uses are; and second, to learn the
scientific method of logical thinking. The first of these reasons is better
served by the treatment which we are adopting, that from the modern
point of view. In this way the facts and theories of chemistry can be
presented to the student in the most systematic and reliable way, and
learned most readily by him; and in this way he can most easily become

familiar with the language of the science and with its present modes of thought.

The second reason, to learn the scientific method of logical thinking, is well served both by the treatment from the modern point of view and by that of the historical development of the science. Indeed, the keenness of logical argument displayed in the early days of the development of theoretical chemistry has rarely been equaled, and the study of the history of chemistry may aid the student to learn the ways of thought involved in the scientific method.

In addition, the history of science is a very important part of the cultural history of the world, and a knowledge of it is needed by the educated man.[1]

"A good understanding of chemistry is a necessity or a help in the practice of nearly every profession—medicine, engineering, geology, physics, biology, running a house," Pauling maintained. His depiction here of what chemists themselves may do is taken from the classic first edition of *General Chemistry,* **published by W. H. Freeman in 1947, two years after the end of World War II.**

A man or woman who selects chemistry as a profession . . . has many roads open to him—he may become a teacher, and at the same time work to discover something new, to bring deeper understanding into the science; he may be a research man, working either with inorganic substances or with organic substances, with metals or with drugs; he may help to control great industrial processes, and to develop new ones; he may collaborate with medical workers in the control of disease. Even if he selects a profession other than chemistry, he may find himself using his chemical knowledge not only in his everyday work but also in overcoming unexpected problems. . . .

Twenty years ago a young organic chemist began work, for an industrial firm, on the problem of making, out of ordinary chemical substances, giant fibrous molecules which could be used instead of silk fibers or other animal or vegetables fibers. He succeeded in synthesizing nylon—which is not just a substitute for silk, but is a far more useful and versatile material.

The director of chemical research for the Bell Laboratories, Dr. R. R. Williams, whose main work was on problems such as the improvement of insulating materials for submarine cables, devoted his spare time to the isolation of a substance present in very small amount in rice polishings. As a young man he had seen, in the Philippines, the ravages of the dietary-deficiency disease beriberi, caused by a diet consisting largely of polished rice and prevented by unpolished rice. By great effort he and his collaborators isolated a gram of the active substance. By the methods of chemical analysis and organic chemistry they succeeded in determining the arrangement of the atoms in the molecule of the substance, and then in finding a way of synthesizing it cheaply from ordinary chemicals. This substance is thiamine, vitamin B_1; its use during the past decade has brought health to people suffering from a deficiency of the vitamin.

The improvements that chemistry has made in metals and alloys, other structural materials such as plastics, and the many other materials such as oils which are used in the machines of our mechanical civilization have been so numerous that they cannot be listed. One may be mentioned as an example—the discovery of a special lubricating oil which does not get thin and useless in hot weather, or thick and sluggish in freezing weather, but which retains a constant viscosity, independent of temperature. This oil is a compound of silicon; it is called silicone oil. Other oils become more fluid as the temperature is raised, because at the higher temperature the molecules undergo more violent thermal agitation, and slide over one another more easily. Silicone oil consists of long, thin, flexible molecules, which, however, at low temperatures are nicely coiled up, each looking like a compact little sphere; and because these spherical molecules can roll over one another rather easily the oil is not very viscous, even though the temperature is low. As the temperature is raised the molecules begin to uncoil, and to get tangled up with one another, in such a way as to interfere with their motion past one another just enough to balance the effect of the more violent molecular agitation accompanying the rise in temperature, and thus to keep the viscosity of the oil constant.

Chemistry has always been of great value to medicine. In recent years we have seen the discovery of the sulfa drugs and of penicillin, which have largely overcome the danger of the infectious diseases. The degenerative diseases—cancer, heart disease, kidney disease, cerebral hemorrhage—are now the most important diseases of death, and present an imposing challenge to the medical-research worker. Our knowledge of the structure of the cells and molecules which make up the human body is not yet great enough to provide an understanding of what these degenerative diseases really are, and to suggest effective methods of attack on them. But chemistry and physics and biology have been developing so rapidly in recent years, and the methods of investigation have been becoming so powerful, that we may look forward confidently to great progress in the future in the understanding and control of these diseases, and to the further alleviation of human suffering. There is need for able and imaginative chemists in the attack on this problem, as well as for medically trained research men with a sound knowledge and appreciation of chemistry.[2]

In *College Chemistry,* the second textbook Freeman published, in 1950, Pauling described the chemist's vast province of concern.

THE universe is composed of matter and radiant energy. The chemist is primarily interested in matter, but he must also study radiant energy —light, X rays, radio waves—in its interaction with substances. For example, he may be interested in the color of substances, which is produced by their absorption of light.

Matter consists of all the materials around us—gases, liquids, solids. This statement is not really a definition. The dictionary states that matter is "that of which a physical object is composed; material." Then it defines material and physical object as matter, so we are back where we started. The best course that we can follow is to say that no one really knows how to define matter, but that we agree to start out by using the word. Often in science it is necessary to begin with some undefined words.

All matter has mass. Chemists are interested in the masses of materials, because they want to know how much material they need to use to prepare a certain amount of a product. The mass of an object is the quantity that measures its resistance to change in its state of rest or motion. The mass of an object also determines its weight, which is only a measure of the force with which the object is attracted by the earth. This force depends on the mass of the object, the mass of the earth, and the position of the object on the earth's surface, especially the distance of the object from the center of the earth. . . . It is common practice to refer to the mass of an object as its weight, although they may not be identical.[3]

As for matter itself, for new students just opening and perusing his textbook Pauling broke it down into different categories. Like all good teachers of introductory courses, he was attentive to the need to explain basic information simply, patiently, and in the proper succession.

AS we look about us we see material objects such as a stone wall or a table. . . . The chemist is primarily interested not in the objects themselves, but in the kinds of matter of which they are composed. He is interested in wood as a material (a kind of matter), whether it is used for making a table or a chair. He is interested in granite, whether it is in a stone wall or in some other object. Indeed, his interest is primarily in those properties (characteristic qualities) of a material that are independent of the objects containing it.

The word *material* is used in referring to any kind of matter, whether homogeneous or heterogeneous.

A *homogeneous* material is a material with the same properties throughout.

A *heterogeneous* material consists of parts with different properties.

Wood, with soft and hard rings alternating, is obviously a heterogeneous material, as is also granite, in which grains of three different species of matter (the minerals quartz, mica, and feldspar) can be seen.

Heterogeneous materials are mixtures of two or more homogeneous materials. For example, each of the three minerals quartz, mica, and feldspar that constitute the rock granite is a homogeneous material.

Let us now define the words substance and solution.

A *substance* is a homogeneous material with definite chemical composition.

A *solution* is a homogeneous material that does not have a definite composition.

Pure salt, pure sugar, pure iron, pure copper, pure sulfur, pure water, pure oxygen, and pure hydrogen are representative substances. Quartz is also a substance.

On the other hand, a solution of sugar in water is not a substance according to this definition: it is, to be sure, homogeneous, but it does not satisfy the second part of the above definition, inasmuch as its composition is not definite, but is widely variable, being determined by the amount of sugar that happens to have been dissolved in a given amount of water. Gasoline is not a pure substance; it is a solution of several substances.

Sometimes the word "substance" is used in a broader sense, essentially as the equivalent to material. Chemists usually restrict the use of the word in the way given by the definition above. The chemist's usage of the word "substance" may be indicated by using the phrase "pure substance." . . .

Substances are classified as elementary substances or compounds.

A substance that can be decomposed into two or more substances is a *compound*.

A substance that cannot be decomposed is an *elementary substance* (or *element*).

Salt can be decomposed by an electric current into two substances, sodium and chlorine. Hence salt is a compound.

Water can be decomposed by electric current into two substances, hydrogen and oxygen. Hence water is a compound.

Mercuric oxide can be decomposed by heat, to form mercury and oxygen. Hence mercuric oxide is a compound.

No one has ever succeeded in decomposing sodium, chlorine, hydrogen, oxygen, or mercury into other substances. Hence these five substances are accepted as elementary substances (elements).

At the present time [over 100] elements are known. Several hundred thousand compounds of these elements have been found in nature or made in the laboratory.

The process of decomposing a compound into two or more simpler substances is sometimes called analysis. The reverse process, of forming a substance by combining two or more substances, is called synthesis.

The composition of a compound can be determined by analysis. For example, a qualitative analysis of salt might be carried out by decomposing it with an electric current and identifying the products as sodium and chlorine; the chemist could then say that the salt is a compound of the two elements sodium and chlorine. To carry out a quantitative analysis he would have to weigh the substances; he could then report the composition as 39.4 percent sodium, 60.6 percent chlorine.[4]

The transformation of matter that takes place in chemical reactions— such as the dramatic one with sugar, potassium chlorate, and sulfuric acid that instantly converted thirteen-year-old Linus into an aspiring chemist—occurs because atoms of two or more elements, separate or in compounds, somehow undergo rearrangements of the bonds among them, some bonds breaking and new bonds forming, yielding new molecules. Aside from the chemical nature of the atoms themselves, which is of overriding importance, factors such as temperature, pressure, electric potential (as in batteries), the concentration of substances in solution, and the introduction of catalysts generally affect the occurrence and rate of chemical reactions. Pauling would often emphasize the fascinating changeability of matter, as he did here when talking to a lay audience at UC Berkeley in 1983. (When lecturing freshmen students, he was more likely to prove some point like this by a dramatic demonstration.)

SUBSTANCES change into other substances. When iron is in contact with air or water, it forms iron oxide or iron hydroxide—what we call rust. The eighteenth-century French chemist Lavoisier proved that pure crystallized carbon—diamond—when combined with oxygen produces carbon dioxide, CO_2. His contemporary, English chemist Henry Cavendish, got an explosion when he put the gases hydrogen and oxygen together and inserted a flame. But he also got water, H_2O.[5]

Almost everyone knows that "H_2O" stands for water, but not quite so many realize that it also means that each molecule of water consists of two atoms of hydrogen and one atom of oxygen. Someone studying chemistry soon needs to know the one- or two-character symbols, or abbreviations, of the elements' names. They appear in all formulas for chemical compounds and in equations showing reactions between substances that transform them into other substances.

THESE symbols are usually the initial letters of the names, plus another letter when necessary. In some cases the initial letters of Latin names are used. Fe for iron (*ferrum*), Cu for copper (*cuprum*), Ag for silver (*argentum*), Au for gold (*aurum*), Hg for mercury (*hydrargyrum*). K for potassium (*kalium*), and W for tungsten (*wolfram*) are from German. The system of chemical symbols was proposed by the great Swedish chemist Baron Jöns J. Berzelius (1779–1848) in 1811.[6]

If the true molecular structure of a substance is known, it is proper to indicate it in the formula. Hydrogen peroxide, with two hydrogen atoms and two oxygen atoms in its molecule, should be written H_2O_2, and not HO; and cyanuric triazide should be written C_3N_{12} and not CN_4, because each molecule contains three carbon atoms and twelve nitrogen atoms.[7]

Pauling appreciated the mathematical precision of chemistry, in which the numbers of atoms in elements and compounds (molecules) remain constant when a reaction takes place; the complete chemical equation describes and accounts for all atoms and any energy generated from the reaction. It all made perfect sense—whereas too often human thought processes did not.

He used the discipline of chemistry to describe the scientific approach, involving inductive and rational thinking, and to explain the meaning of such words as "hypothesis," "theory," and "law." He believed that the training and practice of science, not the ruminations of philosophy, properly prepared people to solve real problems—whether they concerned the perplexing behavior of some aspect of the natural world or the behavior of people in society.

SCIENTISTS do their work in many ways. A great scientific discovery is often the result of a great flight of the imagination—a brilliant new idea. If you have studied physics, you probably read that Archimedes is said to have been taking a bath when he had his brilliant idea, a "flash of genius," about the change in weight of a body immersed in a liquid (Archimedes' principle). Curiosity and an active imagination are great assets to a scientist.

No one knows the method for having brilliant new ideas, and this is not part of what is ordinarily called the scientific method. But scientists also work by applying common sense, reliable methods of reasoning, to the problems that they are attacking, and the procedure that they follow, which is called the scientific method, can be learned.

Part of the scientific method is the requirement that the investigator be willing to accept all of the facts. He must not be prejudiced; prejudice might keep him from giving proper consideration to some of the facts or to some of the logical arguments involved in applying the scientific method, and in this way keep him from getting the right answer. If you were to say, "I have made up my mind—don't confuse me with a lot of facts," you would not be applying the scientific method.

The remaining part of the scientific method consists of logical argument.

The first step of the scientific method is to obtain some facts, by observation and experiment. The next step is to classify and correlate the facts by general statements. If a general statement is simple in form it may be called a *law of nature*. If it is more complex it is called a *theory*. Both laws of nature and theories are called *principles*.[8]

Pauling introduced students and readers to the importance of hypotheses, theories, and laws to progress in chemistry and in science, and therefore to the scientific approach. In the past four centuries scientists had transformed Western civilization's ideas about the planet and the universe—and also about the human ability ultimately to transform matter itself.

CHEMISTRY deals with the properties of many thousands of substances. As the years go by many new substances are discovered in nature or are made in the laboratory, and many new phenomena are observed. These new discoveries and observations make the science of chemistry broader; nevertheless the science has not become harder to learn in recent years. It has, indeed, become easier, because the facts of chemistry are being continually correlated and systematized by theories.

When it is first found that a picture or a mathematical equation correlates or explains a number of empirical facts, the picture or equation is accepted as a *hypothesis*, to be subjected to further tests and experimental checking of deductions that may be made from it. If it continues to agree with the results of experiments, the hypothesis is dignified by the name of *theory* or *law*.

Theories and laws are of great value in simplifying science. A knowledge of Newton's laws of motion, for example, is a knowledge of the behavior of countless mechanical systems, such as a falling ball, a moving projectile, the solar system, a derrick or similar machine. This knowledge permits reliable predictions to be made of how a system will behave, even before it has been constructed or thoroughly studied and observed in detail. The theories and laws of chemistry are not so precise and exact as those of physics. However, this very lack of perfection may account for the intrinsic interest chemistry can have for the student—there is still great room for discovery in chemistry. Already, however, the theories and laws of chemistry have been discovered to be of the greatest value in simplifying and ordering the immense body of chemical knowledge.[9]

In his own century, Pauling became known as the theorist who most contributed to organizing information and then proposing theories that would simplify understanding the laws of chemistry, particularly with regard to the structure of molecules and the bonds within them that joined the atoms together.

To explain the importance of some theory or discovery in chemistry, Pauling often led listeners or readers through a brief review of its origin and development, as he did with the very notion of the atom—the basic unit of matter as viewed by chemists.

THE concept of atoms is very old. The Greek philosopher Democritus (ca. 460–370 B.C.), who had adopted some of his ideas from earlier philosophers, stated that the universe is composed of void (vacuum) and atoms. The atoms were considered to be everlasting and indivisible—absolutely small, so small that their size could not be diminished. He considered the atoms of different substances, such as water and iron, to be fundamentally the same, but to differ in some superficial way; atoms of water, being smooth and round, could roll over one another, whereas atoms of iron, being rough and jagged, would cling together to form a solid body.

The atomic theory of Democritus was pure speculation, and was much too general to be useful. . . .

In the early nineteenth century, however, came a hypothesis that explained many facts in a simple and reasonable way. It became the most important of all chemical theories. In 1805 the English chemist and physicist John Dalton (1766–1844), of Manchester, stated that all substances consist of small particles of matter, of several different kinds, corresponding to the different elements. He called these particles atoms, from the Greek word *atomos*, meaning "indivisible." This hypothesis gave a simple explanation or picture of previously observed but unsatisfactorily explained relations among the weights of substances taking part in chemical reactions with one another. As it was verified by further work in chemistry and physics, Dalton's atomic hypothesis became the atomic theory.[10]

"By the time I read about this theory, around 1915, the existence of atoms was not yet accepted as a fact," Pauling commented elsewhere when describing the history of more recent discoveries in chemistry and physics in the context of his own boyhood knowledge.

IN a popular textbook of the time atoms were defined as the "imaginary units of which bodies are aggregates." The article on "Atom" in the eleventh edition of the *Encyclopaedia Britannica*, published in 1910, ends with the words "The atomic theory has been of priceless value to chemists, but it has more than once happened in the history of science that a hypothesis, after having been useful in the discovery and coordination of knowledge, has been abandoned and replaced by one more in harmony with later discoveries. Some distinguished chemists have thought that this fate may be awaiting the atomic theory. . . . But modern discoveries in radioactivity are in favor of the existence of the atom, although they lead to the belief that the atom is not so eternal and unchangeable a thing as Dalton and his predecessors had imagined." Now, only half a century later, we have precise knowledge of the structure and properties of atoms and molecules. They can no longer be considered "imaginary." The existence of atoms is now accepted as a fact, and nuclear physics continues to divide them and identify smaller and smaller entities within them.[11]

In his surveys of the history of chemistry, Pauling liked to introduce names, dates, and contributions—and show the consequences of a concept or a discovery.

THE term "chemical structure" was used for the first time in 1861, by Butlerow, who stated that it is essential to express the structure by a single formula, which should show how each atom is linked to other atoms in the molecule of the substance. He stated clearly that all properties of a substance are determined by its molecular structure, and suggested that it should be possible to find the correct structural formula of a substance by studying the ways in which it can be synthesized.

The simple ideas about valence and the chemical bond that were proposed one hundred years ago have been of inestimable value to science and to mankind. They grew, slowly but steadily, into the present chemical structure theory, which may be described as one of the most inclusive and powerful generalizations ever made by man about the nature of the universe.[12]

Valence established the exact combining abilities of many elements. But what holds these atoms and molecules together within the basic elements and the innumerable compounds comprising them?

IN the 1860s, about fifty years before I went away to college, chemists in Germany, England, and France had decided that the atoms in substances generally can be described as forming bonds with one another. It was accepted that the hydrogen atom can form one bond, the oxygen atom can form two bonds, the carbon atom can form four bonds, and the silicon atom can form four bonds. For fifty years after 1865 chemists had made great progress in understanding the properties of substances by discussing the various ways in which atoms can be attached to one another by these chemical bonds.

It had been found that most of the substances in living organisms are compounds of carbon. For this reason, compounds of carbon began to be called organic compounds, and this usage has continued. Some remarkable ideas had been formulated. First, it was suggested that the four bonds formed by a carbon atom are directed in space toward the corners of a regular or nearly regular tetrahedron. This idea, this idea of a tetrahedral carbon atom, explained in a reasonable way some of the remarkable properties of certain organic compounds, and it was accepted by most chemists. Chemists were able to assign structural formulas to the molecules of many organic compounds, showing the atoms as circles or symbols connected by lines representing the chemical bonds. Chemists, especially organic chemists, make much use of these chemical formulas in planning and interpreting their experimental work, and as a result the science of chemistry progressed very rapidly.[13]

However, with inorganic substances that lacked the certitude of carbon's tetrahedral linkages to atoms of other elements, structural chemists could never be sure that the theoretical bonds they proposed for com-

pounds were correct. They needed some laboratory means of determining the precise way in which atomic bonding in molecules took place. They became increasingly interested in the different geometrical structures that many elements and compounds formed, and by the early 1900s both physicists and chemists recognized that crystals conveniently provided clues to the basic structures of the atomic bonds holding molecules together.

In 1922 Linus Pauling, already interested in the chemical bond, would find himself on the ground floor center of this research going on in various laboratories: in Europe, notably in England and Germany, and in the United States primarily at the California Institute of Technology. Considering his intense interest in atomic structure and theoretical science, he said later that he might just have readily become a physicist.

BUT to me chemistry is, in some ways, a far more interesting subject than physics because it deals with the real world in what seems to be a good "real" way. For instance, a physicist may be interested in the properties of metals in general. A chemist is interested in the specific properties of lithium, sodium, magnesium, potassium, calcium, titanium, and so on—for all seventy-four different metals.[14]

And Pauling, though a master theoretician, always loved the specific ways that matter is manifested in the universe. At Caltech he would commence investigations that led him from mineral crystals to the particular structural shapes of hundreds of molecules, and finally to the intricate biochemistry of blood, protein tissues, and brain.

Three

Education in Science

Education, true education . . . is preparation both for a life of appreciation of the world and for a life of service to the world.

Education—one's own continuous pursuit of knowledge and the parallel sharing of this knowledge with others—was fundamental to Linus Pauling. If there was a profession that Linus Pauling would have ranked alongside that of the scientist, it would be teaching. With him, the two roles were inseparable. Each was, or should be, concerned with forming a strong link between two persons, the mentor-expert and the protégé-novice, and also between generations—the earlier one informed by years of experience resulting in wisdom, the later possessing the mental and physical energy of youth. The strength of this instructional and inspirational connection meant progress in knowledge, technology, the arts, and morality in any society. Public education, the preparation of a future citizenry, also determined a society's very survival—which is probably why Pauling the scientist and educator began veering in the last half of his life toward a third role: that of the prophet-reformer, who contrasted how things were with how they *should* be.

Linus Pauling knew well the value of teachers and mentors. Deprived of a father when he was nine years old, he had always found satisfactory guides and role models as he grew to maturity, especially among teachers who recognized his potential and encouraged the development of his intellect. The protégé did not disappoint their expectations. And though he outgrew their limitations, he always remembered them in detail, fondly and vividly. After accomplishing impressive discoveries and achieving renown, he spoke with great gratitude of those who had been his mentors in his elementary and high schools, in college, and in graduate school.

Pauling often spoke and wrote about education. In 1951, when both he and the world were at midcentury, he was winding down after a yearlong tenure as president of the American Chemical Society (ACS), the large national organization of professional chemists that Pauling had joined as a student. Accustomed by then to being a prominent spokesman for science, Pauling proposed a solution to individual citizens' ignorance as well as the irrational blundering of humankind. Toward the end of the year he published "The Significance of Chemistry to Man in the Modern World" in Caltech's monthly magazine, *Engineering and Science.* Its ambitious subtitle, "A proposed program for the education of the citizen in science, which would start at the kindergarten level," telegraphed his notion of improving American citizenry by putting a coherent science education into every school system in the nation.

While reading the excerpts from this long article, keep in mind that over four decades have gone by since Pauling wrote it.

I T is impossible to deny that science has played a major part in determining the nature of the modern world. The food that we eat, the clothes that we wear, the means of transportation that we use in going from place to place, the medicines that keep us well, the weapons that we use in killing each other have all been changed in recent years through scientific discovery.

It may well be contended that the world is now in a dangerous situation because science and its applications have developed faster than the understanding of the average citizen. It is evidently of great importance to attempt to improve this situation through a program of education of the citizen in science. The world in modern times has continued to move toward the ideal democratic system, in which all important decisions are made by the people as a whole. In order for this system to operate correctly the citizen must have knowledge enough of the world to make the right decisions; and in the modern world this means that the citizen must have a significant understanding of science. . . .

It is of individual importance to the citizen in the modern world to have an increased understanding of the scientific aspects of the world. The citizen who is trained in science may also be expected to exercise his political rights more effectively than one not trained in science, both because of his greater understanding of the nature of the modern world and because of his understanding of the scientific method, the way in which conclusions can be drawn from facts. Understanding of the scientific method confers on him the scientific attitude, which gives him an increased chance of reaching the right decision about political and social questions as well as about scientific questions. . . .

A citizen who understands the nature of scientific generalizations will ask himself what the basis is for generalizations in the field of economics, politics, international relations. How many facts have been used in determining the attitude of one nation toward another nation? Is the number of facts in agreement with one another in supporting the attitude so great that there is no room whatever for skepticism about the national policy on the part of an individual citizen, or is the citizen justified in being skeptical? . . .

The laws of probability have as much significance in nonscientific fields as in scientific fields. We all understand how a life insurance company calculates its premium payments. The statistician for the life insurance company collects information about the number of people dying at various ages—30, 31, 32, 40, 50, 60, 70, 80, 90, 100, even 110 years of age. From all of these data he calculates the average expectancy of life, and thus finds the average number of premiums that will be paid. If he were to say, "It is abnormal for a man to live to be eighty-five years old, or older; therefore I must discard the data about these abnormal people," he would obtain a wrong value about the average expectancy of life, and the life insurance company would go bankrupt. Yet, just this foolish procedure is sometimes carried out in the operation of a democratic system of government. The principle upon which a true democratic system operates is that no single man is wise enough to make the correct decisions about the very complex problems that arise, and that the correct decisions are to be made by the process of averaging the opinions of all of the citizens in the democracy. These opinions will correspond to a probability distribution curve, extending from far on the left to far on the right. If, now, we say that all of the opinions that extend too far to the right—beyond the point corresponding to the age eighty-five in the above example, say—are abnormal, and are to be excluded in taking the average, then the average that we obtain will be the wrong one. An understanding of the laws of probability would accordingly make it evident to the citizen that the operation of the democratic system requires that everyone have the right to express his opinion about political questions, no matter what the opinion might be. . . .

How can the citizen get scientific knowledge? The answer to this question can be drawn from past experience. Hundreds of years ago it was recognized that mathematics is of great value to the individual. Mathematics is a difficult subject; one might be tempted to say that, since it is difficult, the study of it should be put off till the college years. Yet, through experience, we have learned that the way to teach mathematics is to start with the teaching of numbers in kindergarten, arithmetic in the first grade and other elementary grades, and to continue steadily, without interruption, through algebra, geometry, trigonometry, calculus. This is the way in which science should be taught.

The time has now come for the study of science to be made a part of the curriculum in every grade, at every level. There should be a class in science in the kindergarten, in the first grade, in the second grade, in the third grade, and so on. Every boy and girl who finishes grammar school should know science, in the same way that he now knows arithmetic, languages, and history. Every boy and girl who finishes high school should know still more about science. Every college student should begin his college work with a sound knowledge of the whole of science—comparable to the knowledge that he now has, at this stage, of mathematics—in order that he might devote his years in college to the more advanced aspects of the subject. Only in this way can we train citizens for life in the modern world. Only in this way can we develop a citizenry able to solve the great social and political problems that confront the world. . . .

If one classroom period per day were devoted to science throughout the years of primary school instruction, the work might begin with very simple discussions of the physical world, including simple demonstrations. The fields of knowledge covered would be largely descriptive in nature, in all branches of science—physics, chemistry, biology, geology, astronomy, geography, etc.; but in addition there could be, even in the earliest years, instruction in the scientific method and the scientific attitude.

The concepts of chemical change and of atomic structure and other concepts of modern science are no more difficult to understand than the concept that the earth is round. We teach students in the elementary schools that the earth is round, even though convincing proof is so difficult that the fact has been generally accepted only during the last few centuries. In the same way the important basic principles of atomic science could now be taught to beginning students, in the elementary schools, with rigorous proof of the truth of the concepts deferred until a later time.

The principal practical problem accompanying the introduction of instruction in science for one class period each day in all school grades is that of deciding what activities the instruction in science should replace. This decision is not an easy one to make, and presumably the subjects to be replaced or on which decreased emphasis is to be laid would be somewhat different in different schools and in different parts of the world.

There has been just as great reluctance to introduce extensive teaching of science into the field of adult education as into the field of elementary education. The best methods to be used in giving scientific information and instruction in the scientific method to mature individuals who have only an extremely limited background in science probably still need to be discovered. It is likely that a thorough study of existing alternative methods and a search for new ones would yield very important results.

The argument might be presented that it is hopeless to attempt to give the average citizen an understanding of science, because of the complexity of science at the present time and the enormous rate of increase in scientific knowledge. How can even the foremost scientist keep abreast of the rapidly advancing front of knowledge when millions of new facts are being discovered every year? I believe that this pessimism is not justified, and that, indeed, science as a whole is becoming simpler, rather than more difficult. Many parts of physics have already passed through the stage of greatest complexity—the stage at which the body of knowledge in the field consists of an aggregate of largely disconnected facts. With the recognition of relationships among these facts, great numbers of them can be encompassed within a single principle. An understanding of the field as a whole can then be obtained by the process of understanding the general principles. It is not necessary for every fact to be learned; instead a few of the facts can be considered, in order to discover their relationship to the general principles, and thereafter other facts that, for practical or accidental reasons, come to the attention of the individual can immediately be correlated by him with the general principles. . . .

One way in which an increased knowledge of the nature of the physical and biological world can be of value to the individual citizen is through the conferring on him of an increased equanimity, an increased confidence in natural law and order. The well-being of an individual may be greatly impaired by his fear of the unknown, which may far exceed the fear that he would have of a known danger, which he might prepare to meet in a rational way. In a world in which human beings have achieved extreme powers of destruction of one another, through the use of an astounding new source of energy, the nonunderstanding individual might well become extremely apprehensive, in such a way as to prevent him from making the correct personal and political decisions, and to cause him to accede without a trace of protest to suggestions or orders from a dictatorial leader.

An incidental advantageous result of scientific education for all people which is of more than negligible significance is the personal satisfaction and pleasure that accompany pure knowledge and understanding. The physical and biological world in which we live is truly astounding and wonderful. No matter what the extent of his general education or the caliber of his mental abilities, every human being might achieve satisfaction and increased happiness through an increased knowledge and understanding of the world. The sources of happiness in life are not so bountiful that mankind can afford to neglect such an important one. If the ever-present oppressing danger of world war can ultimately be averted, and the world can enter into a continuing period of peace and friendliness, the intellectual activities of the average man may become a source of happiness to him comparable to that provided by his emotions.[1]

In 1950 Pauling wrote a long article, "Academic Research as a Career," for the ACS monthly journal, *Chemical and Engineering News*. More than four decades later, his practical advice may prove particularly helpful to young scientists pondering their future. It may also interest other students wondering whether they should try to remain in academe, venture out into the great marketplace of commerce and industry, or train for some different form of human service.

A career of academic research—research and teaching in a university or college—is the best of all possible careers, for those people who are suited to it by nature and disposition.

By academic research is meant research in pure science, research that is carried out in an effort to increase our understanding of the nature of the world in which we live. It does not have as its primary purpose the solution of any practical problem; but it is fundamental to practical progress, in that the discoveries made by the workers in pure science of one generation provide the basis for the great industries of the next generation.

The overwhelming majority of the workers in academic research are connected with universities or colleges. A smaller number are in research institutes. . . . Some of the great industrial research laboratories provide the opportunity for a few men to carry on fundamental research. Most of the people who carry on academic research also devote part of their time to teaching.

The first question is: what are the qualities that determine whether a student is or is not fitted for a life of academic research? There is one all-important requirement—that he have a deep curiosity about nature, a consuming desire to know more about the world; in short, that he have the scientific spirit. He must be a scholar by disposition. It is good also, if he is to be a teacher as well as a research man, that he have a strong desire to communicate his knowledge to other people.

In addition to scholarly interest, scholarly aptitude is desirable. A brilliant student, with a penetrating mind and a phenomenal memory, has the greatest chance of being a brilliant research man, provided also that he has the scientific spirit, the urge to learn more about nature; but experience has shown us that some men who could not be described as brilliant students have become outstanding figures in academic research. Such a man might be a gifted experimentalist, or a careful, penetrating analyst of fundamental theoretical principles.

It is not necessary that a student be of a conventional nature in order to succeed in academic research. Success may lie in research productivity, or in effective teaching, the training of young men for scientific careers, or in any combination of these two activities; and the positions available for academic research men vary so much in nature that the man with small gifts as a teacher, but with outstanding research ability, and the man with a keen interest in the scientific world and in the instruction of students, and yet with small desire to obtain new knowledge through his own efforts, may both be successfully fitted into academic life. It is indeed true that the popular conception of both the professor and the scientist is that they are freaks; we know that this conception does not hold in general, and that even the academic research man, the combination of professor and scientist, is usually a well-balanced, conventional citizen: and yet there is a place in academic research for the unusual character, and, indeed, we note that it is often the unusual man who makes the most surprising and fundamental discoveries. . . .

The trained research man of highest type has the fundamental traits of originality, ingenuity, and imagination, which cannot be given to a student by graduate training, although they can be developed. . . . In addition he has original ideas, discretion in the choice of what ideas to follow, and the patience and drive to finish the pursuit without jumping from one project to another. All of these qualities are important to the academic research man.

A young man who enters academic research should develop his own *field of research* as early in his career as is possible. He should strive to work independently; one of the most attractive features of the academic life is the freedom of investigational activity that the academic man has. In general, the academic research man in a university has the right to make decisions about the nature of the research that he will carry on, and the way in which the problems will be attacked. In recent years it has become necessary to have large and expensive pieces of apparatus in order to attack certain problems. A member of a team operating a complex apparatus in the attack on a difficult problem does not have the freedom of decision about his research that an independent worker has. It is my opinion that in general this sort of engineering teamwork is not suited to universities, and that the academic research man should avoid the fields of work of this sort in which he would have to sacrifice his independence. . . .

The goal of the academic research man should be to become the authority in a field of knowledge. . . . On the other hand, to limit the field in an extremely narrow way would be undesirable, if it restricted the breadth of his interests. I believe that the best general solution that can be found to this problem is the following: the academic research man should be broad in his interests, and should, in case the facilities available to him permit it, carry on an investigational program along several lines—he should have more than one interest; and in addition, he should devote a considerable fraction of his efforts to one field, perhaps a rather narrow field, of which he should become, after some time, the recognized master.

The primary reason for a man to go into a life of academic teaching and research is that it provides him with an opportunity for carrying on researches in pure science in the fields in which he himself is interested. There are other advantages to academic life that might well appeal to him. The vacation periods are in general much longer than those for employees in industrial or governmental work, and he may use these periods for following up a second interest—another field of science, or even a different scholarly field. Moreover, the academic life offers an opportunity for writing books, which may be scientific monographs, textbooks of science, or even nontechnical books. Finally, academic research offers greater opportunities for travel than are offered by most other careers. It is important for the research man to travel, in order to find out how problems are being tackled in other universities and research laboratories, to learn more about the world generally, and to get acquainted with the other people who are making contributions to science. Herodotus said that it is the scholars who should be travelers— and this statement is just as valid now as two thousand years ago.

The principal disadvantage of a life of academic research is that the income received by a man in this field is significantly less than that which he would receive in other fields of activity. The academic man may increase his salary somewhat through royalties from the sale of books, or through work as consultant for industrial firms. However, he should not plan to obtain a great amount of income from consulting work, because the time devoted to the work might become so large as to interfere seriously with his primary career, of teaching and research in pure science. He should in particular be on guard against the danger of becoming involved in jobs of applied research for industry.

The smaller salary received by the academic research man is such a significant disadvantage that the young chemist or chemical engineer may well ask what possible justification there could be for his decision to follow this career. The answer is that the compensating advantage of freedom to carry on independent research is so great for a man who is deeply imbued with the scientific spirit as to make it worth while to accept the decreased salary.

The training that a student should obtain in preparation for a life of academic teaching and research should be as broad as possible and as fundamental as possible. The student should concentrate on the basic scientific subjects, including mathematics, and on broadening cultural subjects; in general, he should ignore technical subjects of ephemeral value. Languages are important, and the student should learn languages as early in his life as possible, preferably beginning in high school. . . .

Every man who plans to make a career in the academic research field should carry on graduate work leading to the Ph.D. degree. The qualities that are necessary for success in this field should justify his appointment as a graduate assistant or teaching fellow in a university. . . .

During his graduate student period and his early postdoctoral years he should attend as many seminars as possible, in his own and other fields of chemistry and in the related sciences, at which the current literature and new discoveries are discussed.[2]

Pauling felt quite at home among students, especially if they thought or felt intensely along the same lines he did, about some aspect of science or an urgent problem in society. Whatever years he had attained, he could relate to that fiercely ambivalent spirit in youth, who may listen in seeming respect and emulation to august professors, yet all the while think their own thoughts and perhaps eventually defy entrenched authority whenever some injustice is allowed to exist.

In the early 1950s, Pauling, because of his own intractable social and political activism, was in disfavor with the State Department. This resulted in several years' refusal of a permanent passport to travel abroad for scientific conferences. This predicament ended only when he was awarded the Nobel Prize for Chemistry. In Sweden to receive the medal on December 10, 1954, Pauling gave a telling speech to the assembled university students who had carried flags and banners in a torchlight procession honoring the new Nobel Prize recipients.

At whatever stage of life he found himself or whomever he was talking to, Pauling was never at a loss for words of guidance. With nonscientists or the elderly, with individuals or a throng, with friends or strangers, he functioned as a perpetual and perennial teacher, with the bent of a reformer. And frequently, too—as here—he showed a puckish sense of humor that lightened his pedagogic messages.

YOUNG men and women:

On behalf of my colleagues, as well as myself, I thank you for your kind demonstration of friendship and respect.

I am reminded of my own students in California. They are much like you—I have observed that students, young people, are much the same all over the world—and that scientists are the same. There is a world-wide brotherhood of youth and science.

Perhaps, as one of the older generation, I should preach a little sermon to you, but I do not propose to do so. I shall, instead, give you a word of advice about how to behave toward your elders.

When an old and distinguished person speaks to you, listen to him carefully and with respect—*but do not believe him. Never put your trust in anything but your own intellect.* Your elder, no matter whether he has gray hair or has lost his hair, no matter whether he is a Nobel Laureate, *may be wrong.* The world progresses, year by year, century by century, as the members of the younger generation find out what was *wrong* among the things that their elders said. So you must *always* be skeptical—*always think for yourself.*

There are, of course, exceptional circumstances: when you are taking an examination, it is smart to answer the questions not by saying what you think is right, but rather what you think the professor thinks is right. . . .

You will have some great problems to solve—the greatest of all is the problem of war and peace. I believe that this problem has been solved, by the hydrogen bomb—that there will never again be a world war—the knowledge that a world war would mean worldwide destruction, perhaps the end of civilization, will surely now lead to permanent peace.

But it is *your* generation that will have the job of working out the means of preventing disaster, by developing safeguards against paranoiac demagogues who might make nations rabid; *you* will have this great job to do—and I am confident that you can do it.[3]

II

The Structure
of Matter

1922–1954

Four

Probing the Chemical Bond

*It is a wonderful feeling to understand
something about the world that no
one else has ever understood.*

In September 1922, Linus Pauling arrived in southern California to
begin his graduate work at the California Institute of Technology. He
soon discovered that he had found his own place in the universe.

Years later, when reviewing his early years at Caltech, he concluded:

DURING the last forty years I have visited universities all over the
world. I now have the opinion that I had the greatest good luck in having
gone to Pasadena in 1922. I do not think that I could have found better
conditions for a career in physical chemistry anywhere else in the
world.[1]

Pauling's preparation actually began several months before his arrival,
when Arthur Amos Noyes, head of the division of chemistry and chemical engineering, wrote to assign him the task of working out the problems
in the first nine chapters of his newly revised textbook, *Chemical Principles*. Pauling did his homework in the summer evenings after performing
his job as a highway paving supervisor in Oregon.

Noyes had also recommended that Pauling do his doctoral research on
the determination of the structure of crystals, under the supervision of
Roscoe Gilkey Dickinson. Two years earlier, Dickinson became the first
person to receive a Ph.D. from Caltech. He was now the resident crystallographer, an expert in the new Bragg technique, which Noyes had im-
ported from England, of using X-ray diffraction to determine the

structures of crystals. By recording and measuring the many deflected X-ray beams from an X-ray–irradiated crystal, one could—with persistence and sometimes luck, and often with the help of extensive calculation—determine how the atoms in the crystal are arranged, and could thereby find out the lengths of the bonds between the atoms and the magnitudes of the angles between bonds. At that time, many of the crystals studied were minerals. Pauling was a quick learner, and within several months, with Dickinson's help, he deciphered the structure of molybdenite—the first of several hundred crystals whose structure he would happily discover and then publish in years to come, right up to the end of his life.

Pauling's work in crystallography at Caltech provided the necessary key for him to gain entry to the structure of crystals and obtain the knowledge they revealed about the ways in which atoms and molecules bond. In 1992 he wrote this statement for the *Dictionary of Science and Technology:*

CRYSTALLOGRAPHY is the science of crystals. Classical crystallography, the study of the angles between natural plane faces on crystals and of their optical properties, was of great value in chemistry, mineralogy, and other fields of science, especially in the identification of substances. . . .

Modern crystallography started about 1890, when the 230 space groups (combinations of symmetry operations possible for a crystal) were discovered, and developed after 1913–14, when diffraction of X rays by crystals was discovered by Max von Laue and the Bragg equation was formulated by Lawrence Bragg. These discoveries led to much progress in atomic physics through the determination of the wavelengths of X rays, and even in nuclear physics through the analysis of gamma-ray spectra.

Modern chemistry, based largely on knowledge about chemical bonds and the structures of molecules as determined by X-ray diffraction studies of crystals, could not have developed without X-ray crystallography. Molecular biology is also a product of X-ray crystallography. The alpha helix and the pleated sheets of protein molecules and the double helix of DNA were discovered by the aid of the X-ray diffraction patterns, even before the powerful new direct methods of crystallographic analysis of diffraction patterns were developed.

Now these new methods have been applied in the determination of the detailed atomic arrangements of hundreds of protein crystals, giving remarkable insights to the nature of living organisms. Our present great understanding of the nature of the world of atoms, molecules, minerals, and human beings can be attributed in large part to crystallography.[2]

Pauling's working and thinking methods were fine-tuned during his apprenticeship period with Dickinson at Caltech.

THE process of structure determination involved a succession of logical arguments, which were presented to me by Dickinson in a meticulous way, with emphasis on rigor. Dickinson was an especially clearheaded and thoughtful scientist, strongly critical of carelessness and superficiality. I was pleased to learn that questions about the nature of the world could be answered by carefully planned and executed experiments.[3]

Pauling's mentor permanently impressed upon him the importance of these fundamental tactics when attacking scientific problems—which he, in turn, would teach to several generations of students.

PERHAPS the greatest thing Dickinson taught me was how to assess the reliability of my own conclusions. He taught me to ask every time I reached some conclusion: "Have I made some assumption in reaching this conclusion? And what is the assumption? And what are the chances that this assumption is wrong? How reliable is the conclusion?" I have remembered this ever since and have continued to feel grateful to him. If you have an original idea, it is possible to delude yourself into thinking that there are observations that support this idea. Or it is possible that although you think your idea has been developed on the basis of rational argument, somewhere in that argument you have made an unjustified assumption. This checking procedure was important in my development.[4]

At the age of ninety-one, Pauling still vividly recalled that enormously stimulating, formative period in his development as a physical chemist.

I can understand why I became excited, in 1922, when I learned that the techniques of X-ray diffraction by crystals could be used to determine the arrangement of the bonds in a crystal and to give additional information, such as the length of the bonds and the angles between bonds. I was fortunate to be able to make use of this technique, under the instruction of Roscoe Gilkey Dickinson, who had just in 1920 received his Ph.D. from the California Institute of Technology for his work on X-ray crystallography. During my first year as a graduate student I was able, with Dickinson's guidance, to determine the crystal structures of two substances, the mineral molybdenite (molybdenum disulfide) and an intermetallic compound of magnesium and tin. Each of these structure determinations provided valuable information about the chemical bonds and led to suggestions about the relation between structure and properties of substances.

Thus X-ray crystallography provided a great opportunity to make discoveries about the nature of the world.[5]

From the vista of four decades Pauling later portrayed the ambiance of the Caltech campus in the 1920s.

BECAUSE of the small number of graduate students in Pasadena in 1922, it was possible for them to come into close contact with the members of the faculty. For example, together with two or three other graduate students and a faculty member, Ellis, I was taken by Noyes on some camping trips to the Palm Springs region, and all of the graduate students were the guests of Noyes in his oceanside house at Corona del Mar, fifty miles south of Pasadena, where a few years later the California Institute of Technology set up a marine laboratory. These occasions gave opportunity for the unhurried discussion of scientific and practical problems. Dickinson also took me with him on a number of trips to the desert during my first year as a graduate student. Moreover, the members of the staff were not overloaded; in 1922–23 I was the only graduate student being supervised by Dickinson.

In addition to inspiring them with the desire to carry on research, it is necessary in the training of young scientists to give them a good background of the knowledge that has already been obtained. The California Institute of Technology, despite its small size, was already carrying out this function in an admirable way in 1922. Although Noyes, as director of the Gates Chemical Laboratory and a member of the Executive Council, was primarily responsible for the development of the institute as a leading teaching and research school in physical chemistry, it was Richard Chance Tolman who, in my opinion, made the greatest contribution to the achievement of this goal. Tolman, who had come to Pasadena in 1921, had a thorough understanding of the new developments in physics and the conviction that chemical problems could be solved by the application of physical methods. In my first term, the fall of 1922, I found his course Introduction to Mathematical Physics, in which he laid emphasis on the basic principles and their quantitative application to moderately simple problems, to be especially valuable. His courses on advanced thermodynamics and on statistical mechanics applied to physical and chemical problems were outstanding for their clarity and thoroughness. But his greatest contribution to the training of the young

physical chemists in Pasadena came, I believe, through the course called
Seminar in Physical Chemistry. . . .

These courses helped me to overcome the handicap of lack of knowledge of physics and mathematics. The course work of other graduate
students in physical chemistry was similar, but on the average not so
heavy.

I learned a great deal also from the research conferences. The chemistry research conference, held once a week, was made lively by Tolman's
efforts to find out the extent of understanding that had been achieved
by the graduate students. . . . There were two physics research conferences per week; often one was combined with the Physics and Astronomy Club, which met alternately at the institute campus and the
Pasadena laboratories of the Mt. Wilson Observatory.[6]

**Whenever he recollected those three years as a Caltech graduate student,
Pauling invariably would mention the next stage in his professional evolution.**

ALSO, beginning in 1925, a remarkable change occurred in theoretical
physics, through the discovery of quantum mechanics. I was fortunate
to be able to participate not only in the experimental field of X-ray
crystallography but also in the theoretical field of quantum mechanics,
especially in its application to the question of the nature of the chemical
bond. This work kept me and my students busy for about fifteen years.[7]

**After Pauling received his doctorate in June of 1925, he stayed on at
Caltech for a few months as he worked on converting findings in various
crystallographic research projects into more papers for publication. As
had happened before, the canny Noyes had already determined the next
steps of Pauling's destiny—not that Linus was aware of it at the time.**

AFTER many years I have decided that Noyes was quite clever in getting any result that he had decided upon. When I was approaching the
completion of my work for the Ph.D. degree there were many National
Research Council fellows in Pasadena, and I applied for a National
Research Council fellowship, saying in my application that I would go
to Berkeley and work with Gilbert Newton Lewis. There was a rule that
a National Research Council fellow had to leave the institution from
which he had received his doctorate. Noyes had in fact been largely
responsible for setting up these research fellowships of the National

Research Council, and he, of course, had determined what the rules would be.

When I prepared to go to Berkeley, after having received the NRC fellowship, Noyes said, "You have done a large amount of X-ray work that you haven't yet written up for publication. That work could most conveniently be done here in Pasadena, so why don't you just stay here and complete writing the papers?" I had not been thinking much about the future, about where I would get a job after the completion of my fellowship.

After I had held the fellowship for about four months, Noyes said to me, "There are some new fellowships, the first of which were given this year. They are called the John Simon Guggenheim Memorial fellowships. You ought to go to Europe, which is the center of scientific work now, so why don't you apply for a Guggenheim fellowship for the coming year?" I applied for the fellowship, and again my wife and I packed up to go to Berkeley. Noyes then told me that the Guggenheim fellowships are not decided until the end of April, but he said, "You are sure to get a Guggenheim fellowship, so I'll give you enough money to pay the fare to Europe and to support you from the end of March until the beginning of the Guggenheim fellowship. It really isn't worth while for you to move to Berkeley and then make another move to Europe." I agreed, and resigned from my National Research Council fellowship; the secretary of the Fellowship Board sent me a very disapproving letter.

My wife and I went to Munich. After a couple of months, when our funds were getting low, Noyes wrote that I had received a Guggenheim fellowship, and after a year I received word that I had been appointed as assistant professor of theoretical chemistry in Pasadena.

It wasn't until about twenty years later that I realized that Noyes was afraid that I would become a member of the staff at Berkeley if I took up my fellowship in Berkeley, and that he had made his arrangements to keep me from having the experience of being in that interesting department of chemistry. In fact, Gilbert Newton Lewis had made a visit to Pasadena a few months earlier, about the time that I got my doctorate. Many years later I learned that he had come to offer me an appointment, but that Noyes had talked him out of it.

I think that it was probably a good thing that Noyes carried out his machinations. It was a fine experience for me to work in the young and growing institution, the California Institute of Technology, and to have the benefit of a good number of years of contact with Arthur Amos Noyes, as well as with Richard Chase Tolman, Roscoe Gilkey Dickinson, and others there.[8]

Pauling's stay in Europe proved invaluable to his future career, for it enabled him to pursue in depth a new line of research not yet current in the United States: the newly proposed quantum mechanics. Caltech first got wind of this development in physics in the spring of 1925, just as Pauling was completing his Ph.D. requirements. Already familiar with the "old" (1900–1913) quantum theory of Planck and Bohr regarding particles of energy, and conscious of certain shortcomings in it, Pauling surmised that the new quantum mechanics, based on the wave equation of Schrödinger, would have better application to chemical problems and would elucidate the chemical bond. As for Noyes, he saw it made good sense to send Pauling off to the sources of this new physics—and then to bring back to Caltech what he learned. And it would keep him far away for a while from the lure of G. N. Lewis at UC Berkeley. (Lewis was a former student and associate of Noyes at MIT, and they were in fact friendly.)

"My First Five Years in Scientific Research," a reminiscence requested of Linus Pauling in 1992 by the editor of the British scientific journal *Nature* and published posthumously in 1994, summarizes that early period in his life as a chemist, including the European part of it.

WE remained one year in Munich with [Arnold] Sommerfeld and had shorter stays in Copenhagen and Zurich, as well as visits to other centers of research in modern physics.

My education continued during this period, especially during the year in Sommerfeld's Institute of Theoretical Physics. Just as we arrived, in April 1926, Schrödinger had begun publishing his papers on wave mechanics, and Sommerfeld immediately began giving lectures on them. Several other seminars were also devoted to newly published papers in wave mechanics and other aspects of quantum mechanics. I immediately began thinking about my main interest, which was not the hydrogen atom or the hydrogen molecules [initially suggested for investigation], but rather the question of the structure and properties of many-electron atoms and ions and of more complex molecules. My paper was published in *Zeitschrift für Physik* in 1927. This was an interesting combination of old quantum theory and wave mechanics, published earlier by Gregor Wentzel, in 1926.

The treatment involved the use of an atomic model developed in 1920 by Schrödinger, who suggested that atoms and ions containing many electrons might be treated by an idealized model in which the inner shells of electrons were replaced by surface charges of electricity, with a single electron in an orbit penetrating these shells. Wentzel made use of

Sommerfeld's old quantum-theory method of quantizing conditionally periodic systems, with the replacement of the square of the orbital-angular-momentum quantum number by the product of the quantum number and the quantum number plus 1. Wentzel had evaluated the screening constants for X-ray doublets by this theoretical treatment, but his calculated values disagreed with the experimental ones. I noticed that an extension of Wentzel's method could be carried out, which led to reasonably good agreement with experiment. So far as I know, this was the first application of quantum mechanics to atoms and ions containing many electrons.

I then continued to develop this method and applied it to the calculation of theoretical values of the electric polarizability, diamagnetic susceptibility, extension in space of many-electron atoms and ions, and also to the evaluation of a set of values of radii of ions for use in ionic crystals. These papers were published in 1927, as well as a paper giving the results of applying quantum mechanics to the problem of the effect of a magnetic field on the dielectric constant of a diatomic gas.

The fact that I was in Sommerfeld's institute just at the time that Schrödinger published his papers on wave mechanics provides a second example of my astonishing good luck. So far as I know, Sommerfeld's lectures were the first on this subject anywhere in the world. It was somewhat by chance that I arrived in Munich in the spring of 1926; I had in fact thought seriously of going to Niels Bohr's institute in Copenhagen, but circumstances caused me to change my mind. I am confident that this bit of good luck was largely responsible for my having got a rather thorough grounding in wave mechanics within a few months of its discovery, and probably responsible for my later success, during the next five years, in applying quantum mechanics to some previously puzzling questions in the field of chemistry, culminating in the publication of my book *The Nature of the Chemical Bond* in 1939.

There have been other times in my life when I have had good luck and some when I have been less fortunate; but I see no reason to complain. I look back on the five years beginning in September 1922 with much pleasure, and with great gratitude to my remarkable teachers, especially Arthur Amos Noyes, Roscoe Gilkey Dickinson, Richard C. Tolman, and Arnold Sommerfeld.[9]

Quantum mechanics, by resolving many of the inconsistencies and inadequacies that frustrated scientists when trying to apply the old quantum theory to the behavior of atoms and molecules, dramatically altered the ways chemists—with Linus Pauling most prominent among them—ex-

plained the nature of the chemical bond. While in Europe, he met a number of established or up-and-coming scientists his age or even younger; with some he formed professional connections and friendships that would last for a lifetime.

By the autumn of 1927, Pauling was back in Pasadena, ready to assume a position on the Caltech faculty that Noyes, not surprisingly, had offered to him.

ON my return to Pasadena in 1927 as assistant professor of theoretical chemistry, I presented a course on introduction to quantum mechanics, with chemical applications. I gave this course every year for many years, and I also gave it, as well as a course on the nature of the chemical bond, in the University of California in Berkeley, where I spent one or two months each spring for five years, beginning in 1929, as visiting lecturer in chemistry and physics. Other graduate courses that I taught in Pasadena were on the structure of crystals, the nature of the chemical bond, and the theory of the electric and magnetic properties of substances.[10]

In 1985, six decades after quantum-mechanical theory began providing Pauling with the tools he needed to explain chemical bonding in ways that made sense to him and ultimately to other structural chemists, he published a summary, "Why Modern Chemistry Is Quantum Chemistry," in the British journal *New Scientist*. Lay readers unaccustomed to the technical language of physics and chemistry will at least get a glimpse at this arcane realm of specialists; they will also see that Pauling attributed the birth and development of molecular biology to the theoretical approach of physics.

MODERN chemistry took shape during the first thirty years of the twentieth century. Chemistry is a quantum phenomenon, or rather, a great collection of quantum phenomena. Max Planck originated the quantum theory in 1901, and Albert Einstein extended it significantly a few years later. Then, in 1911, Ernest Rutherford discovered the atomic nucleus. This discovery made it possible for Niels Bohr to formulate, in 1913, his theory of the hydrogen atom. Bohr's theory, involving the motion of an electron and a proton in circular orbits about their common center of mass, was a most important contribution to the development of modern chemistry.

Then, in 1916, Arnold Sommerfeld quantized elliptical orbits. This was an important step for chemistry. It permitted chemists, when thinking, for example, about compounds of carbon, to consider that each carbon atom has four elliptical orbits directed toward the corners of a tetrahedron, thus giving a vague quantum significance to the chemist's concept of a tetrahedral carbon atom.

In 1916, Gilbert Newton Lewis, dean of the College of Chemistry in the University of California, Berkeley, proposed that the chemical bond consists at all times of two electrons held jointly by two atoms. Irving Langmuir and others further developed his ideas about the electronic structure of molecules from 1919 on.

In the early 1920s, there was a dispute between Lewis and the physicist Robert A. Millikan of the California Institute of Technology. Millikan supported the dynamic atom, the Bohr atom; Lewis supported the static atom, in which the electrons occupy definite positions about the atom in a molecule, or crystal. The question was: which provides the better picture? We now recognize that the atom is both dynamic and static: the electrons move in the atom, but we can say that electrons, or electron pairs, occupy reasonably distinct regions in space about the nucleus.

Despite the efforts of outstanding physicists, including Wolfgang Pauli and Werner Heisenberg, to develop a treatment of the simplest molecule —the hydrogen molecule ion—using the "old" quantum theory, no progress in the quantum theory of the chemical bond was made until Werner Heisenberg, Erwin Schrödinger, Paul Dirac, Max Born, and others had formulated the "new" quantum mechanics in 1925 and 1926. A young Dane, O. Burrau, applied the theory in 1927 to the calculation of the energy, bond length, and electron distribution for the ground state of H_2^+. His results were in complete agreement with experiment. Edward U. Condon then published an approximate treatment of the electron-pair bond in the hydrogen molecule by placing two electrons in the Burrau ground-state orbital. In the same year, 1927, Walter Heitler and Fritz London developed their treatment of the hydrogen molecule in which the wave function is composed of the wave functions for normal-state hydrogen atoms combined together, with the two electrons changing places about the two nuclei.

Other investigators soon obtained improved results by contracting the hydrogen-atom orbital somewhat, by introducing some polarization, and by introducing ionic terms. With more complicated functions, the calculated properties of the hydrogen molecule were soon in complete agreement with the observed values.

The modern theory of chemistry based on quantum mechanics has made continual progress from 1927 to the present time. The development proceeded along two different lines, each of which has its own value. The first way is to solve the Schrödinger equation, with the use of computers, to obtain more and more reliable theoretical values for various properties of molecules and crystals of ever-increasing complexity. The second method, in which I have been especially interested, is to apply the ideas of quantum mechanics, in combination with empirical information about the structure and properties of substances, to develop an extended semiempirical theory of structural chemistry that helps all chemists in their thinking and in understanding their observations.

Chemists have by now made hundreds of ab initio quantum-mechanical calculations for the ground state and excited states of molecules and crystals. Usually, they have obtained results in good agreement with experiment. In some cases, the principal conclusion is that quantum mechanics gives the right answer to questions about molecular structure and crystal structure. Sometimes the calculations provide some additional insights into molecular or crystal chemistry. Often the computations predict values for certain properties of substances that have not yet been subjected to experimental investigations, and we can often make reliable predictions about unstable substances, such as, for example, the isomer HNC, which is much less stable than HCN. Chemists are also applying molecular quantum mechanics effectively in their efforts to obtain a greater understanding of the mechanisms of chemical reactions. This very important problem has not yet been completely solved.

One of the quantum concepts most important for chemistry is that of resonance between two structures. Heisenberg introduced the idea in 1926, in his discussion of the energy levels of the helium atom. The basic idea is that sometimes we can describe a state such as the normal state of a system, in terms of wave functions that correspond to two (or more) different structures. If we have two equivalent structures, we can construct a symmetric and an antisymmetric combination. One combination of wave functions will predict a lower energy for the system and this combination corresponds approximately to the wave function for the normal state of the system. The hydrogen molecule ion is an example: we can describe the energy of the chemical bond as the resonance energy between two structures, in one of which the electron occupies a hydrogen-atom orbital around one proton, and in the other it occupies a similar orbital around the other proton. A second example is the Heitler-London interchange of the two electrons in the electron-pair bond of the hydrogen molecule. A third example is the stability of benzene, which resonates between two structures. Around 1930, the

physicist John C. Slater developed some of these ideas and made especially important contributions to atomic physics and to chemistry. In
1929, he formulated a method of writing a general wave function for
atoms and then in the same year he extended it to obtain the wave
function for a molecule in which the atoms are connected by covalent
bonds. (In convalent bonding, electrons are shared among the atoms
rather than transferred completely from one to another as in ionic
bonds.)

Quantum mechanics, together with experimental information about
bond lengths, bond angles, and other properties of molecules and crystals obtained by X-ray diffraction, electron diffraction, and several spec-
troscopic techniques, has permitted modern structural chemistry to
develop to the point where it has contributed to much scientific progress. Some examples are the discovery of the alpha helix and the two
pleated sheets as the principal ways of folding polypeptide chains in
proteins, and the discovery of the double helix of DNA, the genetic
material for living organisms. Molecular biology is now one of the most
rapidly developing fields of science.

It is just, I believe, to say that molecular biology is a product of
quantum mechanics, and that Niels Bohr, by formulating his theory of
the hydrogen atom in 1913, made a crucial contribution to this development. [11]

**After his European stay, Pauling kept trying to understand satisfactorily
how and why atoms are joined together in molecules and crystals, so as
to identify and explain the essence of the chemical bond. Years later,
in his talk to students on "The Challenge of Scientific Discovery," he
remembered back to the first stage of an important insight.**

IN 1928 I was thinking about this problem. The carbon atom has four
valence electrons. The spectrum shows clearly that one electron is different from the other three, and yet the four bonds of the carbon atom
seem to be identical with one another, directed toward the corners of a
regular tetrahedron. I had the idea these electrons, the s electron and
the three p electrons, may be all right for the spectroscopist, but they
can be combined together, hybridized together, into four equivalent
electrons occupying four equivalent tetrahedral orbitals. This idea gives
almost a complete theory of the nature of the chemical bond. I worked
at my desk nearly all that night, so full of excitement that I could hardly
write. It is a wonderful feeling to understand something about the world
that no one else has ever understood. [12]

**Still, Pauling had more work to do in putting his insight into a form
that could be understood and then easily used by other people, even by
himself. After a lifetime of discoveries, the elderly Pauling reflected on
that necessary gestation period between the first insight and its translation into a communicable form.**

SOMEONE asked me not long ago what was the discovery I made that
excited me the most. And I answered that it was the basic discovery
about directed chemical bonds that I made in January of 1931. I had in
fact already published a paper in 1928, two years after I began learning
quantum mechanics, in which I said that by a treatment that I called
resonance theory, I could explain the tetrahedral nature of the bonds of
the carbon atom, and that I would publish details later. But nearly three
years went by before I published the details.

In 1928 I was working with the quantum-mechanical calculations,
which were very complicated mathematically, and I managed to derive
the result that the carbon atom would form four bonds in tetrahedral
directions, but it was so complicated that I thought, "People won't believe it because it is so hard to see through this mass of symbols and
equations and relationships. And perhaps I don't believe it either!" It
was taking a long time for me to simplify the quantum-mechanical questions so that they could very easily be applied to various problems.

Then one day in January of 1931, late in the day, I had this idea: "I
can use a simple but powerful method of simplifying these equations.
Then I can apply these simplified equations to various chemical problems. Not only can I easily derive the tetrahedral bonds around a central
atom, but also octahedral ligation and square planar ligation, which
occur with certain substances. And I can make predictions about relationships between magnetic properties and the arrangement of the
atoms around each other." So I worked, I think, nearly all night, very
excited about applying this idea.

I consider that paper ["The Nature of the Chemical Bond: Application
of results obtained from the quantum mechanics and from a theory of
paramagnetic susceptibility to the structure of molecules"], which was
published on the seventeenth of March, 1931, as my most important
paper, and I believe I am right in saying that this discovery is the one
that developed the greatest feeling of excitement in me.[13]

**Pauling had produced his famous theory of hybrid bond orbitals, which
was recognized as a seminal concept. Later that year Pauling was given
the Langmuir Award as the most outstanding young chemist in the**

United States. Not coincidentally, at the age of thirty he was appointed a full professor at Caltech, and two years later was invited to become a member—the youngest yet—of the highly prestigious National Academy of Sciences, established by Congress in 1863 as a private organization that would provide advice to the government on matters of science and technology. In this greatly productive period begun in the late 1920s, he wrote a succession of papers that dealt with such theories as the shared-electron pair bond, resonance, and the electronegativity scale.

Quantum mechanics, then, had enabled Pauling to explain the bonding phenomenon theoretically in a far more satisfactory way than before. He formulated generalizations regarding the atomic arrangements in crystals with ionic bonding, in which negatively charged electrons, orbiting around the positively charged nucleus, are transferred from one atom to another. "Pauling's Rules" proved of great value in deciphering and interpreting ionic structures, particularly the complex ones of many silicate minerals. Pauling discovered that in many cases the type of bonding—whether ionic or covalent (formed by a sharing of electrons between bonded atoms)—could be determined from a substance's magnetic properties. He also established an electronegativity scale of the elements for use in bonds of an intermediate character (having both ionic and covalent bonding); the smaller the difference in electronegativity between two atoms, the more the bond between them approaches a purely covalent bond. To explain covalent bonding, Pauling introduced two major new concepts, based on quantum mechanics: bond-orbital hybridization and bond resonance.

Hybridization reorganizes an atom's electron cloud so that some electrons assume positions favorable for bonding. Since the carbon atom can form four bonds, tetrahedrally arranged—a central structural feature of organic chemistry—Pauling's explanation of it and many related features of covalent bonding got great attention from chemists in research and education throughout the world. Resonance is a rapid jumping of bonding electrons back and forth between two or more possible positions in a bond network. Resonance makes a major contribution to the structural geometry and stability of many substances, such as benzene or graphite, for which a static, nonresonating bond system would be inadequate. Pauling later extended this bond resonance concept to a theory of bonding in metals and intermetallic compounds.

Pauling's innovative concepts, together with numerous examples of their application to particular chemical compounds or compound groups, now gave chemists fundamental principles to apply to the growing body of chemical knowledge. They could also accurately predict new compounds and chemical reactions on a theoretical basis that was far more satisfactory than the straight empiricism of pre-Pauling chemistry. His ideas became incorporated into the standard lexicon of physical chemistry.

Pauling's interest in molecular structure would continue throughout his long career, and the calculations involved gave him both recreation and happiness. So did model building, creating likely molecular structures using paper, wood, metal, or plastic. He used what he called the "stochastic method," which drew upon his own mental storehouse of information and allowed him to postulate a structure, based on reasoning and theoretical calculation. Detailed laboratory verifications would then usually be carried out by associates—as with most of his research projects. (Many of his discoveries and inventions were then expanded upon and utilized profitably in industry by others. And though in later years he was involved primarily in biomedical research, his curiosity impelled him to identify the intricate structures of a number of minerals, transition metals, intermetallic compounds, and other substances, such as the controversial quasi-crystals. In 1992, for instance, he was awarded a patent for a novel technique of fabricating superconductive materials.)

Pauling's grasp of chemical-bonding possibilities enabled him to make an early prediction that is especially significant in view of current biomedical research recognizing the effect of free radical and antioxidant actions on health and aging. In 1931 Pauling proposed the existence of the superoxide radical (SOR). It is not generally known by researchers that it was Pauling who discovered it theoretically, and in 1979 he detailed that particular history. By then the enzyme superoxide dismutase (SOD), present in the body, was known to disarm the marauding superoxide radical. SOD, Pauling said, was "a vital substance that played a crucial role in the evolution of life. The superoxide radical may well be the principal cause of oxygen toxicity, and superoxide dismutase may be essential for survival." However, the superoxide radical itself may have been the initiator of the first primeval organisms. Pauling concluded:

I still retain an interest in structural chemistry, but in recent years I have also become increasingly interested in biochemistry and medicine, and it is a source of satisfaction to me that the superoxide radical, whose existence was predicted through arguments based upon the theory of quantum mechanics, should have turned out to be important in biology and medicine.[14]

That discovery might be classified as an informed hunch. Pauling saw the value in noticing hunches when pursuing or making discoveries.

I once knew a physical chemist who was interested in what he called hunches. A hunch is a kind of inspiration: a new idea that suddenly occurs in the mind. In a paper he presented to the National Academy of Sciences of the United States, this scientist reported on having asked one hundred chemists about their hunches. Four or five people failed to understand his meaning, but many of the chemists interviewed agreed that they sometimes acted on hunches. . . .

Hunches, or inspirations, come to me often when I have thought about a problem for years and then have suddenly found the answer. This is because I train my subconscious mind to retain and ponder problems. Whenever a new idea enters my head, my subconscious mind asks whether it is related to any of the long-standing problems stored there. If there is a connection, the new material is brought to the attention of my conscious mind. This way of thinking is not unique with me. . . .

Investigators in the field of inspirations, or hunches, have suggested that a person who commands several branches of knowledge transfers something that is well known in one area into other areas. The act of transferal constitutes an inspiration. . . .

Not all the many ideas I have had about science have proved correct; nonetheless, I have discovered many things and have published my discoveries. This is probably because I think more about scientific problems than most other scientists. Another reason is my broad background of knowledge. This scope enables me to transfer facts from the field of physics to chemical problems to which they have never been applied before. This process can be rather straightforward, but some of my ideas have been somewhat more inspirational (I do not mean in the sense that the spirit breathed them into me). My inspirational ideas have come from my great body of knowledge.[15]

In 1943 Pauling produced an article that gave historical examples of the important function of unfettered thinking, with a provocative, rhetorical title: "Imagination in Science?" Its rather breezy style, showing Pauling's enthusiasm for the history of physics as well as chemistry, had probably been enhanced by an editor at the magazine *Tomorrow*.

ISAAC Newton's mind was so unhampered by restraint and convention that he was able—in a burst of inspiration—to tie together the apocryphal falling apple and the earth treading its path about the sun. He was able to identify the force which pulls the apple toward the earth with that which pulls the earth continually toward the sun; and in so doing, he started off a whole new scientific era.

The scientist, if he is to be more than a plodding gatherer of bits of information, needs to exercise an active imagination. The scientists of the past whom we now recognize as great are those who were gifted with transcendental imaginative powers, and the part played by the imaginative faculty in his daily life is at least as important for the scientist as it is for the worker in any other field—much more important than for most.

A good scientist thinks logically and accurately when conditions call for logical and accurate thinking—but so does any other good worker when he has a sufficient number of well-founded facts to serve as the basis for the accurate, logical induction of generalizations and the subsequent deduction of consequences.

From the careful observations of many astronomers who labored faithfully at their instruments—especially those of Tycho Brahe—Johannes Kepler drew the inductions regarding the orbital motion of the planets about the sun which are expressed by Kepler's laws. In the formulation of his laws of planetary motion, Kepler was without doubt aided by his imagination. The apparent motion of a planet as seen from the earth—with planet and earth each describing independently its elliptical orbit about the sun—is so complex that it would have been very difficult for the analysis of the data to have been made by straightforward methods. However, once the imaginative picture of the solar system in terms of elliptical orbits had appeared to Kepler—with his imagination stirred no doubt by the well-known interesting properties of the ellipse —its test by comparison with the observational data would have become an easier, though still a most laborious, task.

Kepler's flight of fancy was as outstripped by Newton's as is the hedge-hopping flutter of a hen by the soaring arc of an eagle. The simple law of universal gravitation, the mutual attraction of every pair of material particles in the universe, eluded lesser men than Newton, men whose minds were confused by the multiplicity of observed physical phenomena or whose imaginations were enfeebled by disuse.

Newton's great feat of the imagination consisted not alone in his formulation of the law of universal gravitation: he had, in addition, to find the mathematical laws which govern the motion of a body under the influence of impressed forces, and even to invent a new kind of mathematics, the calculus, in order to make suitable comparison between his postulates and the observed behavior of the physical world.

We have now become so accustomed to the wonders of nature that it is hard for us to appreciate how great a feat of the imagination is the concept of action at a distance displayed in the attraction of the sun for the earth over the intervening space of ninety-two million miles. The difficulty of accepting this concept, let alone of originating it, is so forbidding that for more than two centuries scientists clung to the idea of a medium, the ether, with the aid of which the gravitational forces were thought to be exerted. The quietus was given this idea by another giant, Einstein, whose imaginative powers in our era are comparable with those of Newton in his.

The Irish Fitzgerald and the Dutch Lorentz were Einstein's Kepler. They had shown that the puzzling results of Michelson's and Moreley's experiments on the independence of the velocity of light in the solar system or galaxy could be accounted for by assuming that their apparatus and measuring rods changed their dimensions as their velocity through the ether changed. The explanation of this puzzling conclusion from the experimental facts followed from Einstein's amazingly original and revolutionary theory of relativity, which abandoned the idea of the ether, the idea of absolute motion, the idea of an invariant time-scale, and other ideas which held earthbound all but the young, active, uninhibited imagination of the twenty-five-year-old Albert Einstein.

Closeness of observation and care in investigating even very small indications of something new are essential in good scientific work. The discovery of the positive electron by Carl Anderson illustrates this truism. Anderson was photographing the cloud-chamber tracks produced by cosmic ray secondary particles in a Wilson chamber placed between the poles of a magnet. The heavy particles, atomic nuclei with positive electrical charges, produced dense tracks bent by the magnetic field to the right, whereas the light electrons, with negative electrical charges, made fainter tracks bent to the left. On one of his very many photographs, Anderson saw a faint track bent to the right. Instead of ignoring this, he looked carefully into the technique, eliminated the chance of error, and finally postulated the existence of a new kind of matter, the positive electron, or positron.

A most striking example of an imaginative idea which has had great scientific and practical consequences is that of the cyclotron. Its giant magnet causes the ions in the split cylindrical electrodes between the pole pieces to describe semicircular orbits of increasing size with such increasing speed as to arrive at the region where the accelerating electrical impulse is effective at just the right time to experience its effect. The concept of the cyclotron must have burst with blinding brilliance in the brain of Ernest Lawrence, though he could hardly have hoped that it would lead to as great advances in the treatment of disease and in the study of the structure of the atomic nucleus as have been made with its aid in the eight years since its discovery.

A large part of the science of chemistry and of chemical industry deals with the so-called aromatic substances, substances which are related to benzene. Benzene was discovered by Michael Faraday and, before many years, was known to contain in its molecule six carbon atoms and six hydrogen atoms. Progress in aromatic chemistry became rapid only after the discovery of the manner in which these atoms are bonded together. This was done by Kekulé, who, in a moment of inspiration after a long period of unsuccessful analysis of the facts known about the chemical properties of benzene, saw that these facts are all compatible with a structure of a completely new type: a hexagonal structure in which the six carbon atoms are bonded together in the structure of a ring.

Fritz Pregl of Vienna was imaginative. Twenty years ago the determination of the chemical nature of substances with physiological activity was greatly hampered because rather large quantities were necessary for a satisfactory chemical analysis: the job of isolating this much was often effectively insuperable—vitamins and hormones, present and active in extremely small amounts, are cases in point. Pregl dreamed of a chemical laboratory on a micro scale—beakers and flasks and pipettes and burettes containing single drops of chemical solutions instead of cupfuls, balances a thousand times more sensitive than those which the chemist was accustomed to use. He developed his imagined technique of microanalysis, and thus provided the world with a chemical tool which has made possible many of the recent chemical discoveries in biology and medicine as well as many discoveries in pure chemistry and in chemical technology.

Credit for Charles Darwin's great contribution to knowledge must be given to his imagination as well as to his industry and perseverance. That part of the great mass of observational material in the field of biology which finds its ready explanation only in the theory of the evolution of species was waiting for interpretation long before Darwin's time, and his own additions to it, while important, were not essential.

The development of the theory of quantum mechanics, the discovery of vitamins, the mapping of genes in chromosomes, the determination of the constitution of the stars, the planning and the execution of methods of manufacture of synthetic textiles and plastics—in short, all of the scientific and technical advances which have been made during the present century—are due more to the imagination of scientists than to any other attribute. Each surprising new scientific development of the accelerating sequence serves to emphasize anew the great importance of the imagination in scientific work.[16]

The populist Pauling who liked to address lay audiences was emerging by the mid-1930s. Occasionally, for instance, he delivered a Friday evening demonstration lecture at Caltech. The public was welcomed onto the campus to be educated in science in a less intensely academic way than in the labs and classrooms there. Pauling's interest in oratory actually went back to Oregon Agricultural College. The two years he had participated in debates and public speaking, combined with his teaching experience there, helped to develop his flair for delivering stirring talks. For years, his informative yet flamboyant freshman-chemistry lectures made a highpoint in Caltech's required curriculum.

In 1937 Pauling went to Cornell University for a semester as the George Fisher Baker Nonresident Professor of Chemistry, an honorific position entailing a semester in Ithaca, New York. In October Pauling delivered his first formal lecture to assembled faculty, students, and friends of the university. It was an opportunity to show what structural chemists do and how they think. Pauling began reading his handwritten text, "The Significance of Structural Chemistry."

I have seen on looking over the books published by earlier Baker lecturers that in most cases the lecturer has chosen for the topic of his introductory address some philosophical, economic, or political question. After much thought I have abandoned the idea of doing this, in part because I have been unable to think of any such topic to which I feel I could make a significant contribution. Remembering, however, the statement of Aristotle, "Old men should be politicians, young men mathematicians," I have decided to speak about the subject of structural chemistry, to which I have devoted most of my professional attention for fifteen years.

In his *Mathematical Theory of Relativity*, Eddington wrote, "The investigation of the external world in physics is a quest for *structure* rather than *substance*." This is true in the main not only for physicists but for all scientific investigators. I am taking the risk of boring you with familiar ideas to make a survey of the dimensions of the universe. On a logarithmic scale, there is shown plotted the dimensions with which humans deal. Scientists endeavor to discuss the objects in each region in terms of entities in the region below. Thus the astronomer discusses the universe as a whole in terms of island universes, these in terms of stars and solar systems; the earth is then discussed in terms of rocks and other objects of ordinary dimensions, and these in terms of molecules; the chemist discusses the structure of molecules in terms of atoms, the physicists the structure of atoms in terms of nuclei and of nuclei, now in terms of protons, neutrons, etc. . . .

We see that the large magnitudes, studied by the astronomer, are, while interesting, not of practical importance; whereas, contrary to cosmology, the researches of physics and chemistry in the region of small magnitudes are of great practical significance, as they deal with the structure of our environs.

It is with the region 10^{-7}–10^{-8} cm that we are now concerned. This is now the region of structural chemistry, including in its scope relatively simple molecules.[17]

[Here is Pauling's textbook explanation of the much-used term "Ångstrom" (usually written as Å) in crystallography and nuclear physics:

THE symbol Å stands for a unit of length, the Ångstrom unit, used in measurement of very minute lengths. It is named after the Swedish physicist A. J. Ångström (1814–1874). One Å is 1×10^{-8} cm; that is, one cm is 10^8 Å, or 1 cm = 100,000,000 Å. (Note the meaning of a negative exponent: $1 \times 10^{-8} = 1/10^8$.)[18]

People who still have difficulty with translating metric measurements into the familiar English system may need to be reminded that a centimeter is 0.39 inch.]

The structures of diamond and graphite form a good example of the significance of structural chemistry as developed during the last century and the present one. Each of these substances is made of carbon atoms only. In diamond each atom is bonded to four neighbors which surround it tetrahedrally—this exemplifies the quadrivalence of carbon as suggested originally by Kekulé and Couper one hundred years ago, and also the tetrahedral nature of the atom postulated by van't Hoff and Le Bel fifty years ago. The great hardness of the diamond is due not only to the strength of the carbon-carbon bond, but also to the arrangement of the bonds. The same bonds occur in soft graphite, which owes its softness to the ease with which the layers slip over one another; and also in linen fibers (or other such fibers), which show a tensile strength of several hundred thousand pounds per square inch, as great as that of the strongest steel.

All hard substances, abrasives, such as corundum, carborundum, quartz, etc., are held together by bonds which connect all the atoms in a crystal into one giant molecule in the same way. Quartz, for example, is held together by Si-O-Si-O bonds. . . . We can see the effect of changing the structure—the arrangement of bonds—without changing the nature of the bonds by comparing quartz with mica and asbestos. All of these are essentially the same in composition, but quartz is hard and compact, mica splits into sheets, and asbestos into fibers, because of the difference in which the bonds are arranged.[19]

From 1927 on, Pauling had published numerous papers about various principles he proposed relating to the chemical bond, and by the mid-1930s he had coauthored two scientific books. But he seemed too busy to make a grand, sustained statement in book form regarding the nature of the chemical bond that he was so successfully probing. Visiting Baker scholars, however, were expected to produce a book for Cornell University Press. So in the course of preparing his lectures and assembling his past publications, he found himself writing the book that would bring him his greatest respect and lasting fame among scientists.

Published in 1939, *The Nature of the Chemical Bond and the Structure of Molecules and Crystals: An Introduction to Modern Structural Chemistry* has remained in print for over a half century. It is said to be the most often cited volume in scientific literature produced in the twentieth century. Here is the beginning of Pauling's first chapter, "Resonance and the Chemical Bond."

MOST of the general principles of molecular structure and the nature of the chemical bond were formulated long ago by chemists by induction from the great body of chemical facts. During recent decades these principles have been made more precise and more useful through the application of the powerful experimental methods and theories of modern physics, and some new principles of structural chemistry have also been discovered. As a result structural chemistry has now become significant not only to the various branches of chemistry but also to biology and medicine.

The amount of knowledge of molecular structure and the nature of the chemical bond is now very great. In this book I shall attempt to present only an introduction to the subject, with emphasis on the most important general principles.[20]

Linus Pauling received a great many awards during his lifetime. Usually these became occasions for him to deliver an important lecture. In the summer of 1946, when given the Willard Gibbs Medal by the Chicago section of the American Chemical Society, he talked on the topic "Modern Structural Chemistry," sketching, with sometimes majestic cadence —and several surprisingly lengthy sentences, for Pauling—the history of this branch of chemistry in which he had early chosen to specialize. (His Nobel Prize in Chemistry speech, given eight years later, had the same title and contained some of the same ingredients.)

THE investigations which I have carried on, with the aid of many able collaborators during the past twenty-four years, have covered a broad field of science, including parts of physics, mineralogy, chemistry, and biology; but, though varied in nature, they have had a common feature —an emphasis on structure—and they may all be considered as being comprised in the general subject of modern structural chemistry.

Structural chemistry—the determination of the structure of chemical substances and the explanation of their properties in terms of structure —is an old science, as old as chemistry itself. We remember Lucretius, who two thousand years ago wrote about "honey, with smooth, round molecules which roll easily over the tongue, whereas wormwood and centaury consist of molecules which are hooked and sharp"; and Lomonosov, whose explanations of the properties of solids, liquids, and gases in terms of moving molecules were boldly imaginative two hundred years ago and yet were, we now know, very close to the truth; and Dalton, who first pointed out that the weight relations of chemical reactions have a simple interpretation in terms of the combination of atoms; and Avogadro and Cannizzaro, who showed that even in an element two or more atoms may be bonded together to form a stable molecule; and Franklin, Couper, and Kekulé, who developed the concept of valence—the representation of the combining power of an atom by a small integer; and again Kekulé, with his strikingly simplifying picture of the benzene molecule as a ring of six carbon atoms, each with its attached hydrogen atom; and van't Hoff and Le Bel, with their explanation of the right- and lefthandedness of some substances, first discovered by Pasteur in tartaric acid, as resulting from the tetrahedral arrangement of the four valence bonds of a carbon atom; and Werner, who showed that the spatial arrangement of bonds determines the properties of many substances other than the compounds of carbon, and that an atom such as platinum may, in its quadrivalent state, hold six atoms or groups about it at the corners of an octahedron, or, in its bivalent state, hold four atoms or groups at the corners of a square, rather than a tetrahedron; and we remember, as the last of the great structural chemists of the past, Gilbert Newton Lewis, whose identification, in 1916, of the chemical bond between two atoms with a pair of electrons held jointly by two atoms, vulcanizing them together, initiated the period of theoretical clarification and precise experimental investigation which has in a few decades changed the old, qualitative structural chemistry into modern structural chemistry.

Modern structural chemistry differs from the older science in being precise—quantitative instead of qualitative; lucid instead of vague. . . .

The field of application of modern structural chemistry which seems to me to have the greatest promise for the future is that of the explanation of the physiological activity of chemical substances. Little success has resulted from the efforts of chemists and physiologists to correlate the physiological properties of substances and their ordinary chemical properties, the properties which depend upon breaking the chemical bonds within molecules and forming new chemical bonds. I believe that usually the specific physiological properties of substances are determined not by these strong intramolecular forces, but instead by the weak forces—van der Waals forces, hydrogen bonds—which operate between molecules, and that the understanding of physiological activity will be consequent to the detailed consideration of these forces in relation to the size, shape, and structure of the interacting molecules. Strong evidence for this point of view has already been obtained through the study of the behavior of systems of antibodies, antigens, and haptens; and I am confident that, as our knowledge of the structure not only of simple molecules but also of proteins and other complex constituents of organisms increases, we shall in time achieve an insight into physiological phenomena which will serve as an effective guide in biological and medical research, and will contribute to the solution of such great practical problems as those presented by cancer and cardiovascular disease.[21]

In this last part of the Gibbs Medal speech, Pauling indicated the area with which his structural research had largely been occupied in the previous decade: biochemistry—in its relationship to human physiology and health. During the mid-1930s Pauling expanded his fascination with structural chemistry into investigating the substances of life.

Five

Messages in the Blood

All of these activities of living organisms are
chemical in nature. To understand them we must
understand the substances, how they are made up of
molecules, how the molecules are made up of atoms.

WHAT is life? This is an interesting question, a great question. What is it that distinguishes a living organism, such as a man, some other animal, a plant, from an inanimate object, such as a piece of limestone? We recognize that the plant or animal has several attributes that are not possessed by the rock. The plant or animal has, in general, the power of reproduction—the power of having progeny which are sufficiently similar to itself to be described as belonging to the same species of living organisms. A plant or animal in general has the ability of ingesting foods, subjecting them to chemical reactions which usually involve the release of energy, and secreting some of the substances of the reaction.

All of these activities of living organisms are chemical in nature. To understand them we must understand the substances, how they are made up of molecules, how the molecules are made up of atoms.

Closely related to the question of the nature of life is that of the origin of life. We know that life began on earth about two billion years ago. We do not yet know, with certainty, how it began. The problem of the origin of life is one of the greatest problems that stimulates man's intellectual curiosity.

Many scientists have asked about the possible ways in which atoms could interact with one another so as to give rise to the first spark of life —a molecule with the power of reproducing itself.[1]

Authors and editors often asked Linus Pauling to write prefaces or introductions to their books. Often he would oblige, perhaps in the process expressing his own views or experiences regarding the subject at hand. His comments given above appeared in Irving Adler's *How Life Began,* a book for young readers. They show the transition Pauling the chemist made between his first work, on the structure of inorganic substances such as minerals, and his subsequent entry into biochemistry. The statement also displays Pauling's economy in writing; this passage appeared, with slight variations, in some of his textbooks to introduce the chapter on biochemistry.

Traditionally, chemistry studies the structure of matter and its transformations when reacting to changed circumstances; physics, the behavior of matter and radiant energy. With the arrival of the twentieth century, however, in both physical sciences came an ever-mounting procession of discoveries regarding the fundamental natures and interactions of matter and energy; these were then applied to the other sciences, including biology. No wonder Pauling in his public talks gauged the times of various key occurrences in science to the year of his own birth, 1901. His grasp of science had grown along with the century.

As a structural and theoretical chemist seeking the precise identities of the basic building blocks of matter, Pauling dealt as much with physics as he did with chemistry. When probing the general aspects of the chemical bond and molecular structure, he began by exploring inorganic rather than organic compounds because their more accessible structures could better elucidate the central questions, the nature of chemical bonds and the principles of molecular structures. For many years the precise forms of organic substances usually could only be guessed at empirically, through knowledge of carbon's bonding tendencies. Organic chemistry was therefore primarily a pragmatist's territory (as in industrial chemistry); it had never appealed to Pauling because of its inexactitude. He had actually disliked his class in it at Oregon Agricultural College, and the attitude carried over into his early work at Caltech.

I had been repelled by organic chemistry when I was a student because I couldn't see much sense to it. There were no basic principles except the tetrahedral carbon atom, which is a very important principle. Back when I became a graduate student, it was not known how far apart the atoms are in organic compounds. There was uncertainty about how they are arranged in space. Sometimes wrong chemical formulas were assigned. Roscoe Dickinson . . . in 1922 together with an undergraduate student determined the structure of the first organic compound. . . . There was uncertainty about what this formula was until this structural determination was made. [But finally] it became possible to get a great amount of understanding about organic chemistry. Along in the 1930s I began to think it was a rather interesting part of the world after all, all of these organic compounds.[2]

There was a reason for Pauling's change of attitude.

B Y 1932 I felt reasonably well satisfied with my understanding of inorganic compounds, including such complicated ones as the silicate minerals. The possibility of getting a better understanding also of organic compounds then presented itself. There was as yet not any large amount of experimental information about bond lengths and bond angles in molecules of organic compounds. . . . Structure determinations had been made [of a few] crystals containing molecules of organic substances, but not a great many. In 1930, however, when in Germany, I had learned about a new method of determining the structure of molecules that had been invented by Dr. Herman Mark. It was the electron-diffraction method of studying gas molecules. The determination of the structure of a crystal of an organic compound, even with rather simple molecules, at that time was often difficult because the molecules tended to be packed together in the crystal in a complicated way. The method of electron diffraction of molecules had the advantage that a simple molecule always gave a simple electron-diffraction pattern, so that one could be almost certain of success in determining the structure by this method. Herman Mark had been good enough to say that he was not planning to continue work along this line and that he would be glad to see it done in the California Institute of Technology. He also gave me the drawings showing how the instrument could be constructed. My student Lawrence Brockway began in 1930 to construct the first electron-diffraction apparatus for studying gas molecules that had been built anywhere but in Mark's laboratory in Germany.

Within a few years we and other investigators had amassed a large amount of information about bond lengths and bond angles in organic compounds. This information had great value in permitting new ideas in structural chemistry, such as the theory of resonance, to be checked against experiment and even to be refined. . . . The observations were generally in accord with the results of quantum mechanical calculations, and it became clear by 1935 that a far more extensive, precise, and detailed understanding of organic compounds had been developed than had been available to chemists in the earlier decades.[3]

Many of the substances that my students and I investigated by electron diffraction were organic compounds. My theoretical work had also extended to encompass compounds of carbon. In the early 1930s, I had formulated a quantum-mechanical theory of the tetrahedral carbon atom, extended, in a simplified form, to inorganic complexes. In the course of our determination of the structure of carbon compounds, it was possible for my students and me to investigate molecules such as benzene and to obtain much experimental evidence for the theory of resonance in chemistry.[4]

Pauling's work was drawing him ever closer to investigating life. Biochemistry, the chemistry of biological substances, is a major subdivision of organic chemistry, which involves the analysis and synthesis of carbon-containing substances. And all organisms contain carbon as a major constituent. Pauling noted what these two interrelated branches of chemistry have in common so far as the basic elements were concerned.

ORGANIC chemistry and biochemistry deal largely with compounds of carbon, nitrogen, hydrogen, oxygen, phosphorus, and sulfur, and to some extent other elements, including the transition metals. The hemoglobin molecule, for example, contains four atoms of iron, and various enzymes important to life contain atoms of iron, manganese, cobalt, copper, zinc, molybdenum, or other transition metals. The nature of living organisms can be understood fully only when we have an understanding of the structure of the molecules of which they are composed.[5]

Pauling regarded life, in essence, as a succession of chemical processes that move an organism from conception and growth to reproduction and death—all involving metabolism, which converts nutrients taken in from the outside environment into energy and components for cell differentiation and development, maintenance, and renewal.

Pauling often told the story of how he happened to get into biochemistry and biomedicine in the mid-1930s, altering the direction of his research thus far and moving into a new series of discoveries. Sometimes he briefly explained it in terms of his permanent mind-set.

I'M essentially a theoretical scientist, working on theories. But one sometimes has to go to nature to get the answers to questions, and accordingly I have done experimental work. I determined the structures of scores of minerals, other inorganic compounds, and perhaps one hundred organic compounds by electron diffraction, and I measured the magnetic properties of substances, hemoglobin, for example.[6]

> Pauling's work on hemoglobin was the very start of his entry into biomedical research as well as into the field of what is now known as molecular biology. By moving into biochemistry Pauling was partly responding to his own curiosity, already aroused by the stimulating new presence of biologists at Caltech.
> Pauling frequently retraced for others the evolution of his interest in the molecules of life, as he did in the first chapter of the memoir that he started in 1992 but never continued to completion.

U P to 1929 I accepted as fact the existence of human beings and other living organisms and the astonishing capabilities that these organisms have. So far as I can remember, I did not make any effort to develop a real understanding of the nature of life until 1929. The question that interested me sixty years ago was why there were living organisms of any sort on earth; that is, what was a possible mechanism by means of which life could come about, with its extraordinary capability of reproducing itself, in its many forms, precisely.[7]

> In the lecture "Chemical Bonds in Biology," which he gave in 1983 at UC Berkeley, Pauling set out to discuss "the nature of life, living organisms, and in particular my own efforts to understand life, living organisms, and to satisfy my own curiosity." As a background, Pauling summarized reasons for this shift in interest more than a half century earlier.

M Y interest in the problem of the nature of life began in 1929. It was in that year that the California Insititute of Technology, which had involved mainly chemistry, physics, and engineering, added a new division, the division of biology. Thomas Hunt Morgan was the head of it. He had brought with him from New York (Columbia) several outstandingly able geneticists—Alfred Sturtevant, Bridges, Emerson, and others, and also some of the biologists, notably Albert Tyler in embryology. I soon got interested in what they were doing.

I had been determining the structures of minerals and other inorganic crystals by the X-ray diffraction technique and also applying quantum mechanics to chemical problems, especially the problem of the nature of the chemical bond. The technique of X-ray crystallography was really great. With a little bit of luck it was possible to determine the structure of a crystal in three months and thereby to learn something new about nature—about structural chemistry. I was beginning to feel that the puzzling questions about the world, at any rate about the inorganic part of the world, could be answered. By 1928 I felt pretty well satisfied that the problem of the nature of the chemical bond could be solved. . . .

I learned from the biologists that *biological specificity* is the major problem about understanding life. Biological specificity is the set of characteristics of living organisms or constituents of living organisms of being special or doing something special. Each animal or plant species is special. It differs in some way from all other species. Its outstanding specificity is that of passing on to its descendants its own characteristic nature.

How is it possible for two human beings to have an offspring who is also a human being, and who in fact in many ways resembles his or her parents more than other human beings? Morgan, Sturtevant, Muller, and Bridges had around 1911 developed the theory of the gene, and three of them—Morgan, Sturtevant, and Bridges—were there in Pasadena. I was accordingly stimulated over and over again to ask what the molecular mechanism of this specificity might be.

I doubt that I had real confidence that I would ever understand this problem of the nature of life, but I could not help thinking about it from time to time.

It is not only the organisms that show this remarkable biological specificity, but also parts of them. For example, most enzymes have the remarkable power of being able to make a chemical reaction take place tens of thousands times as fast as in the absence of the enzyme, and the enzymes are very particular about the reaction that is thus catalyzed. Superoxide dismutase catalyzes the destruction of the free radical superoxide. It does not speed up the destruction of the hydroxide radical, singlet oxygen, or anything else, so far as I know.

We manufacture superoxide dismutase. So do all other animal species —but the superoxide dismutase manufactured by other species of animals is different from that manufactured by us. The number of animal species is large—a million, say. But the superoxide dismutase molecule contains thousands of atoms, such that a million changes can be made in it without impairing its enzymatic activity—each animal species can have its own kind of this enzyme (perhaps shared with some very close relatives), as well as its own kind of the thousands of other enzymes and proteins that it makes.

How can this specificity be explained?[8]

Pauling would answer that rhetorical question after diligent searching. In 1934 he was ready to enter the realm of biology. It was a radical career move for someone who had never even taken a college course in biology or biochemistry. However, taking up biochemistry to study the interaction between physiological processes in organisms and chemical substances was not totally an intellectual curiosity; it was also an economic issue.

These were the Depression years. The Rockefeller Foundation had been helping to fund Pauling's studies of the sulfide minerals. In 1934 Warren Weaver, its director, informed Pauling that this support would not continue; they were far more interested in furthering biomedical research. So Pauling realized that research relevant to human health would be more readily and generously fundable than the crystallographic and other structural investigations he had been doing for a dozen years. It was just as well that his intense curiosity about living matter had been aroused by the Caltech biologists. Already intrigued with the unknown biochemical actions of iron molecules within hemoglobin in the red blood cells, he told Warren that he wanted to gain a better understanding of oxygen exchange in the respiratory system by studying the magnetic properties of hemoglobin. Pauling got the grant, and "Warren Weaver then invented the phrase 'molecular biology' for what we were doing."

The investigation would be undertaken with much the same analytical techniques Pauling had used for inorganic matter and also, more recently, with various challenging organic compounds—except that he also arranged to borrow from the Mt. Wilson Observatory a special instrument for measuring magnetism. As with inanimate crystals, in Pauling's first intensive study of a molecule of life he would be looking primarily for structure. Some biological substances, such as that of hemoglobin in blood, conveniently crystallize. His facility in structural chemistry would prove invaluable in directing this and a host of future biomedical investigations.

Whenever Pauling talked about hemoglobin, his words conveyed some of his initial enthusiasm for that first project.

THE red protein in the blood, hemoglobin, contains ten thousand atoms. When I was a graduate student I looked at three big books in the chemistry library at CIT. They were by Schuchert and Brown. These investigators had taken samples of blood from animals of different species, had separated the erythrocytes and hemolyzed them, and let the hemoglobin crystallize out of the solution. There were photographs of the crystals in the books. The crystals were all different—different symmetry, different face development, different interfacial angles, different indices of refraction. Each species of animals makes its own kind of hemoglobin.

Fortunately, I had been attracted by the red color of the beautiful crystals of hemoglobin and had written a paper on the oxygen equilibrium of this problem.[9]

The Rockefeller Institute of Medical Research, affiliated with the Rockefeller Foundation, arranged to send an expert, Alfred Mirsky, from New York to Pasadena to teach Pauling and his associates whatever information and techniques they needed in their investigation. That was the start of a whole new area for Pauling's exploration into proteins.

N O T only did Mirsky teach me how to handle proteins in the laboratory —they are far more delicate than inorganic substances—but he also gave me a great amount of information about the properties of proteins and especially about denaturation of proteins. The result of this collaboration was that within a few months we were able to publish our paper on a theory of the structure of native and denatured proteins. We suggested that the polypeptide chains in native proteins had a well-defined configuration and that they were held in place by interatomic forces, van der Waals forces, interaction of electrically charged groups, and especially formation of hydrogen bonds. We said that an increase in temperature or the addition of substances that would break hydrogen bonds would cause the protein molecule to unfold to some extent, loosening some segments of polypeptide chains that were not held in well-defined configurations, and that the product of this process was the denatured protein. In particular, if the pH of the solution was near the isoelectric point of the protein, these unfolded segments of polypeptide chains would get entangled with one another, fastening the molecules together and ultimately leading to the formation of a coagulum. Moreover, loss of the native configuration would destroy the characteristics of the protein, such as the ability to combine reversibly with oxygen and other molecules, the ability to crystallize, and enzymatic activity. I think that this was the first modern theory of native and denatured proteins.[10]

Having explored the nature of the chemical bond that holds atoms together within molecules and crystals and having sufficiently explained how the basic structural configurations in inorganic matter and some organic compounds occur, Pauling was well equipped to apply his accumulated knowledge to the science of life—to the biological substances

that are or had been in living animals and plants. Life, even in its simplest forms, requires a vast number of different molecules. How are these molecules constructed? And then, how did this structure relate to function? Many molecules, such as the proteins, are macromolecules—giant-sized molecules containing many thousands of atoms. Pauling recognized that the hemoglobin molecule, for instance, "is extremely complex —it has a molecular weight of 68,000, each molecule containing about ten thousand atoms!" [11]

Medical science had previously tended to be pragmatic when seeking solutions to problems relating to human health. But the nineteenth century had seen progress. Louis Pasteur, a biochemist and microbiologist, made great and fundamental contributions to both medicine and chemistry; so had Koch, Ehrlich, and other scientists who used microscopy to discover the microorganisms that caused disease. Pauling could certainly recognize that medical research was more directly beneficial to humanity in relieving suffering and preventing or delaying death than the far more theoretical work he had been doing, which had no immediate or discernible effect on society, however interesting and even valuable it might be to other physical chemists, crystallographers, and mineralogists.

As with so much else in science that Pauling got into, the hemoglobin project led to another, and then on to others that branched out from them in turn. In time, too—in the mid- to late 1940s—Pauling returned to hemoglobin (Chapter 6). In a lecture given in 1983, Pauling articulated how he chose to pursue immunology in the mid-1930s.

I had been working with Charles Coryell, one of my students, on the magnetic and other properties of hemoglobin. When I gave a seminar on that subject at Rockefeller Institute for Medical Research in 1936, Karl Landsteiner asked me to come to his laboratory, which I did the next day. In those days, when it took five days to go from California to New York, one did not go for a day and return the next day; one stayed a week and went to musical comedies, museums, and such during the week. So, I was able to see Landsteiner the next day. He asked me how I would explain the specificity of interaction of antibodies and antigens, and I replied that I could not. [12]

Pauling could not help but feel flattered by Landsteiner's attention.

Dr. Karl Landsteiner, who in 1900 discovered the existence of different blood groups, thus making it possible to carry out blood transfusions with safety, developed a method of study of serological reactions that has provided much information about their nature. He showed that it was possible to cause an animal to manufacture antibodies with the power of specific combination with various chemical groups of known structure. He and his collaborators prepared artificially conjugated antigens by coupling relatively simple substances to proteins, and then injecting these artificial antigens into animals.[13]

From the Caltech biologists he spent time with, Pauling had heard a great deal about what they called biological specificity. Now Karl Landsteiner added more information about this specificity, particularly in its bearings on the immune response in human physiology. And he engaged Pauling in collaborative research.

THIS was a great experience for me—to have this field of knowledge presented and clarified to me by its great master, the man who had contributed more to it and had thought more deeply about it than any other living being; and I have a deep feeling of gratitude to him for his kindness and interest.[14]

One of the interesting things that Pauling learned from his association with Karl Landsteiner was that scientists may approach research studies rather differently.

DURING the time that Landsteiner gave me an education in the field of immunology, I discovered that he and I were thinking about the serologic problem in very different ways. He would ask, "What do these experiments force us to believe about the nature of the world?" I would ask, "What is the most simple and general picture of the world that we can formulate that is not ruled out by these experiments?" I realized that medical and biological investigators were not attacking their problems in the same way that theoretical physicists do, the way I had been in the habit of doing.[15]

The combination of the two approaches seemed to work well. With funding from the Rockefeller Foundation and guidance from Landsteiner, Pauling explored why and how antibodies are rapidly formed in blood serum and lymph by the immune system to counter antigens— invasive substances such as foreign proteins and infectious microbes that cause allergic reactions that sometimes are harmful, even fatal, instead of beneficial. Pauling, who had spent many hours in the past years among

biologists, directed his attention to the crucial mechanism of biological specificity.

Something seemed strangely familiar to Pauling in this work. Later he talked of why this was.

WHEN I was working in the field of immunology, immunochemistry, and in particular studying the important rule of molecular complementariness, it took me some time before I realized that crystals give the best example. A crystal structure of an organic compound is such that, if a molecule is removed, the surrounding part of the crystal has the structure with the greatest complementariness to the molecule that fits into that position. So crystallization is a phenomenon closely similar to gene replication, enzyme activity, and everything else that we think of as biological and involving biological specificity, and of course its specificity is just great. A crystal will reject everything except the molecules that belong in those positions. In the same way an antibody will refuse to combine with haptenic groups except the appropriate ones that are complementary to the combining region of the antibody. . . . There's a lot that might be said to support the statement that crystals are a very important part of the whole universe.[16]

In a later lecture, "Analogies Between Antibodies and Simpler Chemical Substances" (1946), Pauling explained other similarities between the reactions of inorganic chemicals and interactions of living molecules in terms of structural matches that had given him clues to a new discovery.

CHEMISTS are not accustomed to thinking of antibodies as chemical substances, nor of the serological reactions shown by antibodies as chemical reactions. It is true that antibodies, containing tens of thousands of atoms in each molecule, are far more complex than simple chemical substances, with around ten or twenty atoms per molecule; and some immunological reactions, such as the anaphylactic shock shown by a previously sensitized animal when it receives a small injection of the antigenic substance to which it is sensitive, do not have much in common with chemical reactions. Nevertheless, a comparison of antibodies with simple chemical substances and of serological reactions with ordinary chemical reactions provides a basis for understanding antibodies and their behavior, and also suggests new experiments to be carried out.

When an antigen (a substance, usually a protein, foreign to the animal) is injected into an animal, it may cause the production in the bloodstream and cells of the animal of substances which are related in their properties to the antigen which was injected. These substances are called antibodies.

Antibodies are proteins (serum globulin), with very large molecules, the molecular weight being often 160,000 and sometimes as great as 1,000,000. A striking property possessed by most antibodies and antitoxins is that of forming a precipitate with the corresponding antigen. For example, when an animal is injected with egg albumin, antibodies homologous to the egg albumin are formed, and the serum of this animal gives a precipitate when it is mixed with a solution of egg albumin, the precipitate containing both antibody and egg albumin. A most striking property of antibodies is their specificity—the serum homologous to egg albumin may give a very large precipitate when mixed with a solution of egg albumin, and give no precipitate at all when mixed with a solution of any other protein. . . .

The two striking properties of antibodies—the formation of a precipitate, and the great specificity of the reaction between antibody and antigen—have very interesting analogs in the chemistry of simple chemical substances.

After several decades during which the nature of serological precipitation was in doubt, it has now been shown by convincing experiments that a serological precipitate is a framework of antigen and antibody molecules, with molecules of both antigen and antibody in the framework attached to their neighbors by two or more bonds. The molecules of antigen and of antibody are thus to be considered as having a valence (analogous to, but not identical with, ordinary chemical valence) equal to or greater than 2. . . . The mechanism of the formation of a serological precipitate is the following: Each [antibody] molecule attaches to itself two antigen molecules, this process continuing until the framework of molecules reaches macroscopic size, constituting a precipitate. Many of the ordinary precipitates that the chemist meets in his work are formed in the same way. An example is the precipitate of silver cyanide produced when a solution of a silver salt is mixed with a solution containing cyanide ion. The silver ion has the property of forming two covalent bonds with cyanide ion; it attaches to itself two cyanide ions, which stick out on either side of it. The cyanide ion also has the property of forming two covalent bonds, one formed by the carbon atom and one by the nitrogen atom. In this way long chains of alternating bicovalent silver and bicovalent cyanide groups are formed, and ultimately these long chains arrange themselves side by side to form the silver cyanide precipitate.

It is, moreover, of interest that the serological precipitate can be redissolved by addition of an excess of antigen, just as can the silver cyanide precipitate by an excess of cyanide ion; in each system the effect is due to the formation of soluble complexes.

The second property of antibodies which has an analog in the chemistry of simple substances is the great specificity of interaction of antibody and the homologous antigen. It has been shown that this specificity is due to a striking complementariness in structure between a portion of the surface of the antigen molecule and a combining region on the antibody. . . . This complementariness in structure leads to a strong attraction between the antibody molecule and the antigen, because it permits this combining region of the antibody molecule to get into close contact with the antigen molecule; the closer that two molecules can get in contact with one another, the stronger the intermolecular force of attraction between them.

A protein molecule serving as antigen usually has a rather complicated surface structure, and it is easy to see how the great specificity shown by antibodies arises. The reaction shown by simple chemical substances which is analogous to that of specific combination of antigen and antibody is the formation of a crystal of a substance from solution. A crystal of a molecular substance is stable because all of the molecules pile themselves into such a configuration that each molecule is surrounded as closely as possible by other molecules—that is, if a molecule were to be removed from the interior of a crystal, the cavity that it would leave would have very nearly the shape of the molecule itself. We can say that the part of a crystal other than a given molecule is very closely complementary to that molecule. Other molecules, with different shape and structure, would not fit into this cavity nearly so well, and in consequence other molecules in general would not be incorporated in a growing crystal. This is the explanation of the astounding chemical process of purification by crystallization—from a very complicated system, such as, for example, grape jelly, containing hundreds of different kinds of molecules, crystals which are nearly chemically pure may be formed, such as crystals of cream of tartar, potassium hydrogen tartrate.

Although crystallization is the only simple chemical reaction which shows striking similarity to serological reactions with respect to specificity, there are many physiological phenomena which are similarly specific, and for which the specificity can be given a similar explanation. The specificity of the catalytic activity of enzymes is due to a surface configuration of the enzyme such as to make the enzyme complementary to the substrate molecule or, rather, to the substrate molecule in the strained state that occurs during the catalyzed reaction. The specific action of drugs and bactericidal substances has a similar explanation. Even the senses of taste and odor are based upon molecular configuration rather than upon ordinary chemical properties—a molecule which has the same shape as a camphor molecule will smell like camphor even though it may be quite unrelated to camphor chemically. I am convinced that it will be found in the future, as our understanding of physiological phenomena becomes deeper, that the shapes and sizes of molecules are of just as great significance in determining their physiological behavior as are their internal structure and ordinary chemical properties.[17]

From his studies of antigen-antibody interactions, Pauling with Max Delbrück had developed an important and enduring theory that they introduced in their 1940 paper "The Nature of the Intermolecular Forces Operative in Biological Processes." Pauling had become fascinated with the potential of intermolecular fit to explain the fundamental property of life, wherein the exact shape and size of two molecules, one serving as a template for the other, determine destiny. He called this phenomenon molecular complementariness. Pauling's immunological work, which had inspired his insights into biological specificity, continued for five years, until Landsteiner's death in 1943. By then Pauling was convinced that biological specificity, manifested in the molecular complementariness of the immune response, held the key to the explanation of the phenomenon of life itself. He had searched for some secret formula or blueprint that would explain how life originates, differentiates itself genetically by biochemical specificity, grows, reproduces, and eventually ceases to exist—at least in the form in which it had once lived. When DNA's structure was found, this quest would end.

In 1940 Linus Pauling published a notable paper in the *Journal of the American Chemical Society (JACS)*. "A Theory of the Structure and Process of Formation of Antibodies" presented his protein-research-based ideas on what to him had become a prime duality. In the years afterward, Pauling would talk about biological specificity and molecular complementariness as if he and his research associates had truly solved the world's greatest riddle. If he hadn't already attained it, at least with his theory he had come close to the much-coveted "secret of life."

THE one question I have tried to answer is how a human being is able to synthesize macromolecules that are characteristic of him or her. There are clearly two simple alternative ideas about this: One is that a macromolecule with a certain structure may interact with other molecules in such a way as to stabilize the complex of two identical molecules; the other is that a macromolecule strongly interacts with another molecule if it is complementary in structure to the first molecule. This idea of complement pairing of the structure is now in the elementary biology textbooks. As far as I can see, it is just assumed, with no emphasis on the idea, that the basis of biological specificity is molecular complementariness. The fact is ignored that fifty years ago this was not recognized.[18]

By applying his theoretical interests and expertise to the smallest divisions of life, to living cells and also to the chemicals inside and outside that sustained their functions, Pauling was revolutionizing aspects of understanding both biochemistry and biology. He would also influence medical research and its application to human health. Consequently, many scientists consider Pauling the founding father of the new science of molecular biology (though the younger generation is barely cognizant of this history). It now furnishes the means for understanding questions such as how life originated, why life forms change through time, why individuality exists, and how human health may be improved. During the 1930s gaining scientific insights into such issues had seemed a remote possibility, but Pauling was eager to pursue them anyway. Not everything that Pauling proposed as new theories and discoveries ultimately proved to be correct; but much did.

In the late thirties and early forties, the period in which Pauling first produced his theory of antibody formation, the biomedical field of immunology did not have access as yet to knowledge of DNA structure and the whole array of elaborate high-tech equipment (such as electron microscopy) and techniques that evolved for research use in molecular biology. Much of the time biologists could only reason and conjecture about things beyond their range of vision.

For a number of years Pauling's model for antigen-antibody interaction served as a workable hypothesis for expanding immunological research and clinical treatment. He and his colleagues even worked on developing artificial antibodies, as the first step in developing medical tactics for fighting infectious diseases and autoimmune disorders. Then a different explanation was proposed by Frank McFarlane Burnet for the formation of close-fitting antibodies in an immune reaction to an antigen: clonal selection. This theory better satisfied what a new generation of researchers thought, observed, and finally proved. But Pauling at least could still lay claim to the enduring validity of his explanation for biological specificity in antigen-antibody interactions—due to molecular complementariness (usually called complementarity now). In 1990 he wrote:

IN my 1940 paper I provided arguments to support two ideas. One is that a haptenic group of an antigen instructs a folding antibody molecule to fold into a complementary structure. This idea has turned out to be wrong, and instead it is now accepted that antibodies homologous to certain antigens have specific sequences of amino acid residues in them, to determine the specificity, and that the antigen selects a cell that has the ability to produce the homologous antibody and directs it to form a clone.

The other idea is that biological specificity, as exhibited for example in antibody-antigen interactions and also in many other ways, results from a detailed molecular complementariness in structure. This idea has turned out, of course, to be correct, and essentially the whole of modern molecular biology is based upon it.[19]

In his later life, asked to contribute a brief statement to a book on *The Meaning of Life* being assembled by the editors of *Life* magazine, Pauling recalled his early research in biological chemistry and the insights that it gave him, which were philosophical, even spiritual, within the scientific vision—and would partly explain his fierce determination, starting in the 1940s, to save humankind from its own worst qualities.

DURING a period of about a decade, beginning in 1936, my principal research effort was an attack on the problem of the nature of life, which was, I think, successful, in that the experimental studies carried out by my students and me provided very strong evidence that the astonishing specificity characteristic of living organisms, such as an ability to have progeny resembling themselves, is the result of a special interaction between molecules that have mutually complementary structures.

In a world that is not in thermodynamic equilibrium, such as our earth, parts of which are heated by sunlight, it is possible for certain chemical reactions to be favored, for example, by the action of enzymes or other catalysts. A molecule or group of molecules that can catalyze its or their own production is thereby able to prosper. This process, over a period of 4 billion years, has led to the existence of human beings. So we are here, in this wonderful world, with its millions of different kinds of molecules and crystals, the mountains, the plains and the oceans, and the millions of species of plants and animals. We have developed a degree of intelligence that permits us to understand the wonder of the world, and also that has given us the power to destroy the world and the human race.[20]

Pauling's curiosity would always lead him into new terrains of scientific investigation. For some years still, proteins would continue to occupy a great deal of his research attention (Chapter 6). In his biochemistry research Pauling continually experienced new challenges and new routes to discovery. Already aware of how physics and chemistry work together, he was stimulated by the close connection between chemistry and biology. It often became crucial for different scientific disciplines to collaborate in solving difficult problems.

The human capacity to seek and perpetuate an ever greater understanding of the interworkings of matter, radiant energy, heat, and motion, and of the natural world itself, was what produced the technology that civilization was built upon. Thousands of years of observations, experiments, and innovations eventually were formalized into the sciences of physics, chemistry, geology, and biology. Since Pauling believed in finding principles and explanations in science that were as simple and all-encompassing as possible, he also declared that it was important to combine the sciences as much as possible in important investigations, now that each province had established its fundamental body of knowledge.

Not surprising, Pauling viewed chemistry as the most appropriate science to be studied and centrally applied in multidisciplinary work. In 1949 he gave a workshop lecture "The Place of Chemistry in the Integration of the Sciences" and provided some examples from his own work in the past decade and a half in biology and medicine.

In attacking the problem of the integration of the sciences, I feel that I must speak with considerable diffidence, since all that I shall attempt to do is to present a few reflections of my own experience, with the hope that they will be of some general interest.

I think that everyone would agree that it is desirable to have integration in the sciences. By this I do not merely mean that the tools and methods of one science should be available to another. This is not enough. It is true that the use of physical apparatus by chemists, or physiochemical methods by biologists, helps to make progress. You remember that when Kirchoff invented the spectroscope eighty years ago, his friend Bunsen immediately used it to discover cesium and rubidium. And more recently, as soon as the cyclotron was invented chemists in Berkeley and elsewhere began using the artificial radioactive materials it produced in the investigation of chemical reactions. Examples of this kind prove that physical techniques discovered in one branch of science have significant value in the solution of problems in another branch.

There is, however, another type of technique which would be of even more assistance if we could make the transfer from one science to another, and that is the technique of thinking. It is in this area that a great deficiency lies, and I myself am not sure how it may be remedied. The physicists have developed the ability to concentrate in a useful way on the theoretical problems of physics. They have specialists who do nothing but think, leaving to others the jobs of carrying out experiments and collecting useful facts. Every time a new discovery is made, theoretical physicists all over the world cooperate in trying to fit it into our present body of knowledge. This effort to relate new facts to what is already known is not made so intensively in other sciences. Individual investigators in the fields of biology and medicine sometimes make such an effort, but in a rather desultory way, so that the consequences of new discoveries are perhaps never thought out, or at least not for many years. . . .

The objection may be raised that it is difficult to make such transfer in an effective way because of the complexity of our body of knowledge. Teamwork alone cannot solve the problem, for the integration needed is primarily that which takes place in the mind of a single man. Comprehension on such a scale may seem to be a well-nigh impossible task— one that would need the universal genius of a Leonardo da Vinci. Yet I believe that the evolution of science toward simplicity will make integration possible of achievement by ordinary mortals like ourselves.

I suggest that the evolution of a science proceeds through three stages. In the first stage a science is quite simple because it does not encompass very much. This is the stage of amassing facts from experiments: when only a few experiments have been carried out the number of facts established is not too great for one man to comprehend. The individual scientist can then know the whole of his science thoroughly and have a considerable acquaintance with other sciences as well. However, as time goes on and more and more experiments are performed the number of known facts becomes greater, until finally they are so numerous that to master a field becomes a full lifetime's work. It is only after a science has passed through this stage of great complexity that the third and final stage can be reached. I think that it has arrived in physics and is not far away in chemistry. This is the stage of simplification, when in order to master a subject it is not necessary to memorize the results of all the experiments that have been performed, but only to learn the fundamental principles that encompass all of these results.

When a science has reached this stage, it is no longer necessary to test all of its facts by experiment in order to understand it. . . .

I am encouraged to believe that this time is fairly near at hand, and that we shall soon have effective scientists who can cover a broad field. This does not in the least mean that their science will be less sound. I read an article recently in *Science* which stated that a certain university proposed a source of training for broad integrative science. This consisted of one year of general elementary chemistry, one year of biology, one year of mathematics, one year of physics, one year of geology and one year of astronomy, after which the graduate was to go out into industry and act as an integrator—a scientific generalist! Frankly, I think such an idea perfectly horrible. How could a man of that sort be any good? Who would have confidence in him? If we want to have a real scientific generalist, he must be trained by mastering one field of science and then spreading his mastery over other fields. Such a man would, of course, have to be better than average.

Perhaps the next thing to consider is the best field on which to concentrate before the scientist attempts to broaden his area. I feel that it would not be adequate for him to be a master of semantics, for example, or a master of sociology, or even a fine theoretical physicist. What he should do is become a master in the field of chemistry and especially, I might say, of structural chemistry. . . .

It seems to me that theoretical physics would not serve because it does not give a complete understanding of the physical world and the biological world. Physicists in general tend to restrict themselves to the small part of the physical world with which they deal, and to leave out of their studies all such features as the structure and properties of substances in relation to their chemical composition, and the reactions that change one substance into another. . . .

Chemists, however—and biologists also—are interested in the different kinds of matter, not only in the form of substances but also in their more complicated aspects, the objects that are built out of the substances. The idealization the chemist makes is that he is interested in the material substance rather than the object: in, let us say, steel, no matter if it is made into an axe head or into a bottle opener. But the biologist is interested in the shape of particular specimens of protoplasm, whatever protoplasm is. It is the objects themselves that constitute living organisms, and living organisms and their parts interest him. His interest includes chemical composition and physical properties. If the biologist explored the whole field thoroughly, so that he understood the chemical composition, the properties of chemical substances, and the physical properties of chemical substances, he would be a universal scientist himself. But the biologist sometimes is deficient in knowledge of the fields of physics and chemistry. A good chemist, on the other hand, must have a sound knowledge of physics as well as of chemistry, and this general basic understanding of the physical world can be used in expanding in such other directions as are necessary to contribute toward the integration of the sciences. He can move into biology and build upon his knowledge of chemistry the additional biological structure needed for some understanding of the biological field, or he can move elsewhere, as into medical research. It is of course possible for a physicist to go into general science also, if he has first a fairly good understanding of chemistry.

Perhaps I should substantiate my claim that a well-trained chemist—especially a structural chemist—has the best chance of contributing to the integration of the sciences. The structural chemist is primarily interested in how the properties of substances are determined by the way that atoms are arranged together in molecules or larger atomic aggregates. This makes him peculiarly able to contribute with profit to integration. The whole problem of understanding science is, I believe, the problem of relating facts to the concept of structure, first in terms of atoms and then in terms of something still smaller, such as nucleons. . . . It is structure that we look for whenever we try to understand anything. All science is built upon this search: we investigate how the cell is built of reticular material, cytoplasm, chromosomes; how crystals aggregate; how atoms are fastened together; how electrons constitute a chemical bond between atoms. We like to understand, and to explain, observed facts in terms of structure. A chemist who understands why a diamond has certain properties, or why nylon or hemoglobin have other properties, because of the different ways their atoms are arranged, may ask questions that a geologist would not think of formulating, unless he had been similarly trained in this way of thinking about the world. . . .

I should like here to say a word about integration of all the sciences, including the social sciences. Everyone realizes that a closer relationship between the social sciences and the natural sciences would be a wise and splendid thing. . . .

I myself feel that the best way in which the basic sciences can be of help in the solution of our great social and political problems is through the process of education—assisting students to learn what the scientific method is, and encouraging them to use it in their thinking generally. An understanding of physical laws would be useful in attacking social and political questions, and would undoubtedly aid by leading to better decisions in these fields. A physical law is a succinct description of the results of a number of experiments. It is not an inflexible, unchanging dogma. It describes only the experiments that have been carried out up to the time the law is stated. We have Newton's laws of motion, based on such facts as that all objects dropped in a vacuum take the same length of time to reach the ground. Physical laws, however, are always subject to change. Newton's laws turned out to be only approximate after the Michelson-Morley experiment had been carried out. When the experiments that constitute the basis for the theory of relativity had been made, the exact laws were seen to be Einstein's equations of motion. Even those, of course, are not exact laws in the field of the very small, where quantum mechanics is involved, and it appears that something more is needed for exactness in nuclear dimensions.

It seems to me important that the student should understand that these basic laws of nature may, as a result of some new experiment, not be exactly right next year. Trained in such an attitude, the student may not be taken in so easily by bald statements about social and political questions, but may reserve judgment, regard such generalizations as without sufficient basis of evidence, and insist on doing his own thinking. There is always the chance that if we try something else, some new idea, it may prove to be closer to the facts. It is often true, in the field of politics, that the dogma currently in favor is something that has worked well once, and therefore is continually supported, even though it may no longer be useful.

I believe that the greatest contribution that science can make to other disciplines is the inculcation of the scientific attitude—the reliance on fact, not dogma, and the habit of flexibility of mind which this engenders. It is not too much to hope that science may in the long run affect man's nature as well as his habits.[21]

As always, Pauling shared with others his high vision of the proper function of science and scientists in modern society.

Six

Proteins Revealed

I am happy to feel that I now have an understanding

of at least the basic answers to the question

"What is life?"

From the late 1930s until well into the 1950s, Pauling expanded and intensified his interest in the molecular makeup and properties of proteins. He and his research associates were working on protein structure in the 1930s. It was a large, all-encompassing, demanding subject—since proteins abound in life forms. Though crucial to life and health, blood and antibodies—the subjects that initially inspired Pauling's interest in biochemistry—make up only a small fraction of protein substances. Learning more about them, though, had offered Pauling the prospect of more immediate medical applications of discoveries—which particularly interested the Rockefeller Foundation and the closely affiliated Rockefeller Institute of Medical Research.

In his all-out quest now to solve the problem of protein structure, Linus Pauling was becoming a protein specialist. His attack went ever deeper into the new frontier of molecular biology. Many years later, writing "How My Interest in Proteins Developed," Pauling started off, as he often did, with his youth. Also, as frequently happened when Pauling in his old age reflected back on earlier times, he would consider what was known or not known in those days regarding micronutrients, especially the vitamins—which by then (1993) had occupied much of his attention for a quarter century.

M Y first memory about proteins goes back to the spring of 1918, during the First World War, when I was a student in a class on camp cookery

given by the home economics department of Oregon Agricultural College as a contribution to the war effort. Students in the class were, like
me, receiving some military training in the Reserve Officers Training
Corps.

I remember making a loaf of bread and also learning something about
the macronutrients, proteins, carbohydrates, and fats, but nothing
about vitamins—it was too soon after the discovery of vitamins for them
to get mentioned in the course.[1]

**Still backed by research funds from the Rockefeller Foundation, Pauling
sought to identify the structures of particular forms of protein; he looked
also for the overriding principles governing these complex structures in
organisms, just as he had done earlier with the nature of the chemical
bond in molecules of inorganic matter. Proteins, after all, represent the
essence of life. Here is a Pauling textbook summary written in 1947.**

AMONG the most important substances present in plants and animals
are the proteins. Proteins are substances which occur either as separate
molecules, usually with very large molecular weight, ranging from
15,000 to many millions, or as reticular constituents of cells, constituting
their structural framework. Hair, fingernails, skin, tendon, and muscle
fibers consist mainly of protein. The blood contains many different kinds
of protein molecules, in solution in the plasma or within the cells of
the blood. . . . The human body contains many hundreds of different
proteins, which have special structures which permit them to carry out
specific tasks.

All proteins are nitrogenous substances, containing approximately 16
percent nitrogen, together with carbon, hydrogen, oxygen, and often
other elements such as sulfur, phosphorus, iron (four atoms of iron are
present per molecule of hemoglobin), and copper.

When proteins are heated in acidic or basic solutions they undergo
hydrolysis, producing substances called amino acids.[2] . . .

No one has yet succeeded in determining the exact atomic structure
of any protein molecule. Strong evidence has been obtained, however,
to indicate that the amino acids are combined in proteins into long
chains, called *polypeptide chains*. For example, two molecules of glycine
can be condensed together, with elimination of water, to form the double molecule glycylglycine. . . .

The bond formed in this way is called a *peptide bond*. This process
can be continued, forming a long chain containing many amino acid
residues. . . . It is believed that protein fibers such as silk, hair, finger-

nails, and muscle fibers consist of long polypeptide chains of this sort, either stretched out into the zigzag form . . . or coiled back and forth in a more complex zigzag, or spiral. Molecular proteins, such as oval-bumin, insulin, and hemoglobin, are thought to be composed of poly-peptide chains folded back and forth into a compact structure.[3]

When Pauling wrote the early versions of his basic chemistry textbook during the 1940s, he did not say a great deal about proteins. He was still investigating their various forms and could hardly conjecture much as yet about their structures. In 1983, at the University of California, Berkeley, Pauling reminisced to an audience of students, faculty, and other interested listeners about aspects of his early research involvement with proteins, starting in the mid-1930s.

A T that time [1937] . . . little was known about the structure of pro-teins. In fact, there was even some doubt that proteins consist of poly-peptide chains. An English scientist, Dorothy Wrinch, had proposed structures in which there were no polypeptide chains with carbonyl groups with double bonds to the oxygen atoms, but rather single-bonded two-dimensional aggregates of atoms. The distinguished American sci-entist Irving Langmuir even joined her in publishing a paper giving arguments for her "cyclol" theory, which is now forgotten. My colleague Carl Niemann and I felt it necessary to publish a paper in which we presented overwhelming evidence for the polypeptide chain structure, which had been formulated around 1900 by the German chemist Emil Fischer.

By 1937 I had reached the conclusion that I knew enough about carbon atoms, nitrogen atoms, and hydrogen atoms to discover the structure of alpha-keratin (hair, horn, fingernail, muscle). I knew (I thought) the bond lengths and bond angles, and I was sure that the peptide group was planar. . . . Alfred Mirsky and I in 1936 had written that proteins are held in their native conformations by hydrogen bonds. Also, the X-ray photographs of hair and other alpha-keratins seemed to show that the structure was repeated every 5.1 Å along the axis of the hair.

I worked on the problem much of the summer of 1937 and failed to find the structure. I decided that there was something strange about proteins, something that I did not know.

Then Dr. Robert B. Corey came from New York to work with me. He, too, was interested in proteins. We knew that no structure had yet been determined for any crystal of an amino acid, the building blocks of proteins, or any simple peptide. We decided to determine some of these structures.

During the next eleven years a dozen amino-acid structures and several peptide structures were determined in our laboratory—none anywhere else in the world: Many graduate students and postdoctoral people did the work, along with Dr. Corey.[4]

Occasionally, almost cryptically, Pauling would refer to the period in the early to mid-1940s as especially revelatory. In a chronological notation— one of his self-notes, possibly for a lecture presenting milestones in his evolution as a researcher—he handwrote, "1940–1945: when I began to understand life." In this period too came his self-noted "rediscovery of rationalism,"[5] perhaps reaffirming his earlier scientific quest as if to parallel other people's spiritual journeys. Such revelations are apt to take place in midlife, whether precipitated by a crisis or naturally occurring. His discoveries in antibody and protein research gave him some scientific authority in understanding the origins of life as well as the processes that enable its perpetuation. He did not need the Bible's Genesis story to serve as an article of faith: a believer in Darwinian evolution, he had his twin tenets of molecular complementariness and biological specificity, which disclosed to his satisfaction "the secret of life" (presented in Chapter 5).

These busy, difficult, and challenging years for Pauling, beginning notably in 1941, caused him to divert attention away from the protein investigation that most characterized his work in the previous half decade—with two large digressions. One was the threat of a terminal illness; in early 1941 Pauling was diagnosed with glomerulonephritis (Bright's disease). Fortunately he was treated successfully by Dr. Thomas Addis, a renal specialist at Stanford Hospital, who put him on an experimental meatless diet that obtained the essential amino acids for protein synthesis largely from plant foods. He did not eat meat for ten years. Ironically, the kidney disease—probably triggered by an infection—was an autoimmune disorder in which one's own antibodies (Pauling's special research area) attacked some type of body tissue.

The other distraction from his protein research work was the U.S. entry into World War II in December of 1941. The United States government utilized Pauling's services in a number of special ways. The Department of the Navy engaged Pauling, his Caltech laboratories, and a group of about fifty researchers in programs for developing high-yield explosives and sophisticated rocket propellants. He devised invisible inks for writing secret messages, which subsequently could be revealed by applying particular reagents. Pauling was also asked to construct an instrument that would gauge oxygen levels in submarines and airplanes to warn crews when oxygen was too low or too high—perilous to life in either extreme. This made an interesting challenge. He invented an oxygen meter based on his knowledge of the magnetic properties of hemoglobin. Eventually it was developed for commercial use by Arnold Beckman, a Caltech chemist who became an entrepreneur specializing in technological transfers of discoveries into patents. This monitoring device proved greatly beneficial to infants in incubators and to patients in surgery, when maintaining safe oxygen levels is crucial. (Beckman Instruments is now a large corporation that supplies biotechnology equipment. Beckman subsequently gave millions of dollars of his profit-based fortune to Caltech and funded many other educational and research endeavors in chemistry.)

Pauling also served in scientists' advisory groups to the government, notably the National Defense Research Council and the Office of Scientific Research and Development (OSRD). As his nephritis retreated and his customary physical energy returned, he piled on the accomplishments, which obviously enlisted the help of many associates. His wartime work—involving at least a dozen major projects—eventually resulted in his being awarded the Presidential Medal for Merit, in 1948, by President Harry S. Truman. (In after years the statement issued with it became an important document in proving his dedication and loyalty as a U.S. citizen.)

In 1946 Pauling began working on the follow-up to a serendipitous encounter that came from his wartime participation in a committee that was investigating the status of medical research.

ONE day, in New York, after we had interviewed the deans of the medical schools in New York and were sitting at dinner together—the members of the committee—a man from Harvard Medical School, Dr. William Castle, said that he had been working on a disease called sickle-cell anemia, in which the red cells of the blood are twisted out of shape. I did not pay much attention; cells seemed to me to be too complicated for me to be interested in. However, when he said that the cells are twisted out of shape only in the venous circulation, and regain their normal shape in the arterial circulation, I thought immediately: Why is it that this difference between the arterial and venous blood exists? It must be the hemoglobin, because in arterial blood the hemoglobin is oxygenated, and in the venous blood it is not oxygenated. These people must manufacture a different kind of hemoglobin from ordinary people.

They have inherited some genes from their parents; it may be that the father and mother are heterozygotes, each carrying an abnormal, mutated gene that manufactures an abnormal sort of hemoglobin. One-quarter of the children would inherit the abnormal genes from the father and the abnormal gene from the mother and have only abnormal hemoglobin in their red cells.

Let us assume that this abnormal hemoglobin molecule has two sticky patches that are self-complementary, mutually complementary in such a way that the molecules clamp on to one another to form long rods, and the rods line up side by side to make a long needlelike crystal that twists the red cells out of shape; but when the oxygen molecules attach themselves, they constitute bumps that hold the molecules away from one another so that the sticky patches do not come into contact. The normal hemoglobin sticky patches and normal red cells do not undergo the change in shape.[6]

As often happened, Pauling had made an intuitive, immediate mental leap into the solution of a problem or the discovery of some scientific rule or phenomenon. (Of course, as in this instance, he usually had a deep reservoir of research knowledge.) But though certain of the logic in his new theory, as a scientist Pauling would have to prove his hypothesis. It ended up involving the work of three years and various researchers.

WHEN I came back to Pasadena a young man, Harvey Itano, who had an M.D. degree, came to work with me for his Ph.D. degree. He obtained some samples of blood of sickle-cell anemia patients and carried out an experiment by putting the hemoglobin in salt solution between a pair of electrodes; the hemoglobin from these patients moved toward the cathode, showing that the molecules had a positive electric charge, whereas that from other human beings moved toward the anode, showing that they had a negative electric charge.

This is the first time that it had been shown that some human beings manufacture hemoglobin molecules different in structure from those of other human beings, and that these molecules can be responsible for disease. . . . Our studies extended over some years, altogether; but I remember the feeling of excitement when, during the few seconds after my friend, Dr. Castle, had talked about sickle-cell anemia, I thought that it might be possible that this disease is a disease of the hemoglobin molecule rather than of the red cells of the blood.

Diseases of this sort may be called molecular diseases. By learning the molecular structure of the molecules that cause these diseases we can understand what the mechanisms of the diseases are, and possibly may develop drugs on the basis of this knowledge.[7]

In 1949 Pauling and his research-physician colleague, Dr. Harvey Itano, announced in a paper published in *Science,* "Sickle-Cell Anemia, a Molecular Disease," that they had discovered the cause of this hereditary blood disease especially prevalent among people of African descent. It was already known to be genetically transmitted. This first identification of a molecular disease served as a springboard for much of Pauling's thinking in the years to come.

Once the Second World War was over, Pauling and his protein-research team and assistants returned to investigating proteins at a much-accelerated pace. Just as he had looked earlier for unifying principles and rules in the bonds holding inorganic or organic molecules together, Pauling sought to decipher proteins, the highly complex compounds in organisms that consist mostly of unique sequences of amino acids linked together by peptide bonds to form polypeptide chains— units that might contain thousands of atoms. It was known that proteins combine polypeptides such as alpha and beta chains, in various configurations, but nobody as yet has determined a chain's actual shape. Until this was done, proteins' three-dimensional molecular structures would remain riddles.

Still, though ardently and arduously pursuing fundamental protein structures, Pauling found himself thwarted—as he had been in the late 1930s, when he and close associate Robert Corey had first begun to seek polypeptide-chain structures. Initially they did this "without success. . . . I decided then that I didn't have enough data. I thought that there must be something unusual about proteins that I didn't know, so I initiated a program of investigation on proteinlike substances."[8] By the late forties the protein-research team concentrated mainly on exploring the fibrous proteins in such substances as horn, fingernails, hair, and silk, and in defining the crystal structures of the amino acids.

In this period Pauling seized upon the metaphor of architecture to describe the bonding structures of elements and molecules. (Coming from him later would be a primer, beautifully illustrated with colorful drawings by Roger Hayward: *The Architecture of Molecules.* It became popular with anyone who wanted to learn the basics of structural chemistry.) In the article "Molecular Architecture and Biological Reactions," published in *Chemical and Engineering News* in 1946, Pauling explained why structural research was vitally important to science and medicine—and how it could be conducted in the future. Contemporary biochemists and molecular biologists may be interested to read about

the status of knowledge a half century ago, when Pauling was immersed
in this promising research area.

THERE are two subjects that I am deeply interested in—structure, the
detailed structure of molecules, crystals, and cells, described in terms of
their constituent atoms, with interatomic distances determined to within
0.01 Å, an interest that began in my youth and has received most of my
attention until recent years; and the basis of the physiological activity of
substances, an interest that is more recent but just as keen. It is with a
deep feeling of satisfaction that I have reached the firm conclusion in
recent years that these two fields are most intimately related.

Why have we still so little understanding of the structural basis of the
physiological activity of chemical substances, despite the interest and
effort of many able physiologists and chemists during recent decades? I
believe that it is because the problem has been examined, in the main,
from one point of view only—not the wrong point of view, but one
which, unaided, gives a vista insufficient to reveal the true complex
nature. This point of view is that which surveys the chemical activity of
molecules—their tendency to break their chemical bonds, the very
strong kinds between atoms, and to form new chemical bonds. The
other point of view which is needed is that which directs the mind's eye
to the detailed size and shape of the molecules and the nature of the
weak interactions of molecules with other molecules, in particular with
the macromolecules and macromolecular stromatic structures which
characterize the living organism. Until very recently physiologists and
pharmacologists have barely thought of this aspect of their great problem—and I am convinced that once they begin to use this new idea
seriously a period of the greatest development will have started. I believe
that the next twenty years will be as great years for biology and medicine
as the past twenty have been for physics and chemistry.[9]

**During 1948, Pauling, on sabbatical leave from Caltech, was a visiting
professor in residence at Oxford University in England. In May, when
giving in Nottingham the masterful lecture "Molecular Architecture and
the Processes of Life," Pauling summed up the past accomplishments,
present status, and future promise of molecular research. That there
was still much for him and other researchers to discover Pauling made
abundantly clear. Here are excerpts from his talk.**

THE last twenty-five years have seen great progress made in our understanding of the nature of life. This progress has been along two lines:

first the chemical substances that make up the living body have been isolated, and information has been obtained about their properties and about the work that they do, and second, great insight has been obtained into the structure of the molecules of chemical substances generally, in terms of atoms and electrons, and this understanding of the properties of chemical substances in terms of their molecular architecture is now being extended to include the very complicated substances responsible for life.

What are the features that are characteristic of a living organism? As we look about us we see such organisms everywhere—human beings, other animals, plants; and we know that there are very many forms of life that we do not see so easily. . . . They [too] are able to grow in size, and to reproduce themselves, and they hand on to their progeny the specific characters that they themselves possess. . . .

Can we obtain an understanding of these properties? Do we know what the nature of life is? I believe that we can understand these properties of living matter, and that we do know what the nature of life is (aside from consciousness), in terms of molecular architecture, the atomic structure of the molecules that constitute living organisms.

Let us first consider how the body works. It does its work by use of special molecules, molecules that have specific properties suitable to the use to which they are put. A few of these molecules are simple ones, representing ordinary chemical substances, such as water, oxygen, carbon dioxide. Others are more complicated, with ten or twenty or thirty atoms per molecule. These include the necessary food substances called the vitamins—vitamin A, vitamin B, vitamin C, and so on, with formulas such as $C_{20}H_{30}O$ (Vitamin A), and special foods such as sugar, $C_{12}H_{22}O_{11}$. And then there are the giant molecules, containing tens of thousands or hundreds of thousands of atoms, and with certain well-defined properties—the ability to do very special jobs that serve the purposes of the organism. Thus in living we make use of oxygen of the air, to burn certain materials in our tissues, and in this way to obtain heat and energy to keep our bodies warm and to permit us to do work, and also at the same time to get rid of some unwanted materials by burning them to water and carbon dioxide. . . .

Under ordinary circumstances the presence of oxygen molecules, in the air, in the neighborhood of a food or other combustible material does not lead to the oxidation of this material. It is usually necessary to light a fire, in order to raise the temperature high enough for the process of oxidation to go on. The body operates at body temperature, 98.6°F, by having developed some ways of causing this burning to go on even at this relatively low temperature. . . . The body contains many special enzymes connected with the oxidation of the breakdown products of foodstuffs in the body, and with the oxidation of the parts of the body itself that are no longer needed. Just as hemoglobin has the specific ability of combining with oxygen and carrying it from the lungs to the tissues, so do the various enzymes have the specific ability of oxidizing certain materials. In general each enzyme has one use: it catalyzes one reaction of the many thousands of reactions that take place in the living body.

Moreover, it is giant molecules, presumably molecules of nucleoprotein, that determine the characters of individual living organisms and that are involved in the transmission of these characters to their progeny. These giant molecules are the genes, which are usually present in structures in the cell called chromosomes. . . .

I have been especially interested in that aspect of biological specificity that is involved in the mechanism of protection of the body against disease. When we first become infected with the measles virus, we become ill. After a few days, however, the protective mechanism of the body has come into operation, and we recover from the disease— the symptoms of the disease were the result of the multiplication of the molecules of measles virus within our body, and the recovery is the result of the development of a police force that stops this multiplication. This police force consists of special molecules, formed in the cells of the body, and circulating in the blood stream. These molecules are antibodies, antimeasles antibodies, with the specific property of being able to combine with the measles virus and to prevent its reduplication. Other diseases also lead to the development of special antibodies to counteract them, and even foreign proteins generally, whether they are harmful or not, cause this process of production of special antibodies to become operative.

The antibodies are closely related to a protein, normal gamma globulin, which is present to the extent of about one per cent in the blood stream. The molecules of antibody against measles are not, however, identical with the molecules of normal gamma globulin. They have instead a certain definite structure that permits them to combine with the molecules of measles virus. The specificity of antibodies is great— the antibodies that protect us against measles, causing us to be immune to this disease after a first attack, do not protect us against any other disease. In general we must build up an immunity against each disease, either through exposure to the disease, which may cause a mild attack or a severe attack, or by some special process. Inoculation against diph-

theria, typhoid fever, and other diseases, and vaccination against smallpox have in recent decades led to the avoidance of an immense amount
of human suffering that would otherwise have occurred.

The process of vaccination against smallpox consists of the introduction into the body not of the virus of smallpox itself, but instead of a
few molecules that are closely similar in nature to smallpox virus. The
similarity between these molecules, of vaccinia virus, and the molecules
of smallpox virus is so great that the antibodies that are produced in
response to the injection of vaccinia virus molecules have the power
also of combining with smallpox virus, and preventing this virus from
reduplicating itself. This very simple and sensible method of combating
smallpox has reduced it from the terrible scourge that it once was to a
rare disease, that can become important again in the civilized parts of
the world only if we forget that vaccination is necessary to keep the
disease under control. . . .

All atoms attract all other atoms, with the general van der Waals
electronic forces of attraction. These forces are weak in case the atoms
are far from one another, and strong only when they are essentially in
contact. The forces between a few atoms in an antigen molecule and a
few atoms in an antibody molecule would not be enough to produce a
bond between the two molecules strong enough to resist the disrupting
influence of thermal agitation of the molecules. If, however, the combining region of the antibody molecule is complementary in configuration to a portion of the surface of the antigen molecule, so that a large
number of atoms of the antibody molecule are able to bring themselves
into contact with corresponding atoms in the antigen molecule, then
the integrated forces of attraction become large, enough to constitute a
significant bond between the two molecules. Other types of intermolecular interaction—the formation of hydrogen bonds, and the forces of
attraction between a positive charge and a complementary negative
charge—may also contribute significantly, if the structures are complementary with respect to them also. It is clear that a good approximation
of the combining region of the antibody to the surface of the antigen
can be achieved only by having complementary structures, and if the
surface of the antigen were changed by adding onto it a group of atoms
even as much as one or two atomic diameters in size, this change might
effectively prevent a large part of the combining region of the antibody
from getting into satisfactory contact with the surface of the changed
antigen, and thus prevent the formation of a significant bond.

The detailed mechanism by means of which a gene or a virus molecule produces replicas of itself is not yet known. In general the use of a
gene or virus as a template would lead to the formation of a molecule

not with identical structure but with complementary structure. It might happen, of course, that a molecule could be at the same time identical with and complementary to the template on which it is molded. However, this case seems to me to be too unlikely to be valid in general, except in the following way. If the structure that serves as a template (the gene or virus molecule) consists of, say, two parts, which are themselves complementary in structure, then each of these parts can serve as the mold for the production of a replica of the other part, and the complex of two complementary parts thus can serve as the mold for the production of duplicates of itself. In some cases the two complementary parts might be very close together in space, and in other cases more distant from one another—they might constitute individual molecules, able to move about within the cell. . . .

Progress in the understanding of the molecular basis of serological reactions and related biological phenomena in which molecules interact specifically with one another has been very rapid in recent years, and we may hope confidently that a great deepening of our understanding will be obtained as the result of the work of scientists in the near future. This progress will not be of value only in satisfying our curiosity; it will, instead, surely lead to significant practical results, especially in the battle against disease.

During the first half of the present century the workers in the field of medical research have won out in the battle against most of the infectious diseases. The diseases of childhood—diphtheria, measles, scarlet fever, whooping cough—are no longer the scourges they once were, the death rate from them being as low as one-twentieth of those of three decades ago. It is the degenerative diseases—heart disease and related diseases of the kidney and peripherovascular system, and cancer—that now are the leading causes of death of human beings. The problem of attacking these diseases is made most difficult because of the lack of complete understanding of their nature, and even of the nature of the human organism itself, in terms of molecular structure.

When once we know what the molecular architecture of the proteins and other large molecules that carry the physiological activity of the human body is, what the relation of the structure of these molecules is to that of the vectors of disease, and of the drugs, such as penicillin and the sulpha drugs, that serve effectively in protecting us against infectious disease, what changes in molecular architecture are associated with the degenerative diseases—then we can attack the problem of the degenerative diseases in an effective way, using the methods of attack that are suggested by this knowledge.

The study of molecular structure is as important a part of medical research as is the work of the clinical investigator in the hospital. We may have confidence that, through the joint efforts of these research men, working in different fields, further great progress will be made, leading to a great increase in the well-being and happiness of man.[10]

With the war over, Pauling had resumed his previous dedicated attention to protein-structure research and reinforced Corey's investigations. But he felt thwarted by the lack of significant progress in finding the elusive architecture of proteins. His talk at Nottingham, however, showed a certain new exuberance. A few weeks earlier, he felt he had experienced a breakthrough and achieved the basis for a significant discovery. But he told no one except Ava Helen about it.

In 1982 Pauling was asked to write an article for a scientific volume about biochemistry. He did this, but the project did not get published, so, for more than a dozen years, the manuscript for "The Discovery of the Alpha Helix" was filed away at the Linus Pauling Institute. Here is the portion that pertains to the title. The pages leading up to this point summarize the origin and sequence of Pauling's interest in protein investigations. (A small selection, relating to his early work with structural studies of organic compounds, was given in Chapter 4.)

IN the spring of 1948 I was in Oxford, England, serving as George Eastman Professor for the year and as a fellow of Balliol College. I caught cold, and was required to stay in bed for about three days. After two days I had got tired of reading detective stories and science fiction, and I began thinking about the problem of the structure of proteins. By this time Dr. Corey and the other workers back in Pasadena had determined with high reliability and accuracy the structures of a dozen amino acids and simple peptides, by X-ray diffraction. No other structure determinations of substances of this sort had been reported by any other investigators. I realized, on thinking about the structures, that there had been no surprises whatever: every structure conformed to the dimensions—bond lengths and bond angles, and planarity of the peptide group —that I had already formulated in 1937. The N—H $\cdots$ O hydrogen bonds, present in many crystals, were all close to 2.90 Å in length. I thought that I would attack the alpha-keratin problem again.

As I lay there in bed, I had an idea about a new way of attacking the problem. Back in 1937 I had been so impressed by the fact that the amino-acid residues in any position in the polypeptide chain may be of any of twenty different kinds that the idea that with respect to folding they might be nearly equivalent had not occurred to me. I accordingly thought to myself, what would be the consequences of the assumption that all of the amino-acid residues are structurally equivalent, with respect to the folding of the polypeptide chain? I remembered a theorem that had turned up in a course in mathematics that I had attended, with Professor Harry Bateman as the teacher, in Pasadena twenty-five years before. This theorem states that the most general operation that converts an asymmetric object into an equivalent asymmetric object (such as an L amino acid into another molecule of the same L amino acid) is a rotation-translation—that is, a rotation around an axis combined with a translation along the axis—and that repetition of this operation produces a helix. Accordingly the problem became that of taking the polypeptide chain, rotating around the two single bonds to the alpha carbon atoms, with the amounts of rotation being the same from one peptide group to the next, and on and on, keeping the peptide groups planar and with the proper dimensions and searching for a structure in which each NH group performs a 2.90 Å hydrogen bond with a carbonyl group.

I asked my wife to bring me pencil and paper, and a ruler. By sketching a polypeptide chain on a piece of paper and folding it along parallel lines, I succeeded in finding two structures that satisfied the assumptions. One of these structures was the alpha helix, with 4.6 residues per turn, and the other was the gamma helix. (The gamma helix has a hole down its center that is too small to be occupied by other molecules, but large enough to decrease the van der Waals stabilizing interactions, relative to the alpha helix. It seems to me to be a satisfactory structure in every respect [other] than this one, but so far as I am aware it has not been observed in any of the protein structures that have been determined so far, and it has been generally forgotten.)

I then got my wife to bring me my slide rule, so that I could calculate the repeat distance along the fiber axis. The structure does not repeat until after 18 residues in 5 turns, the calculated repeat distance being 27.0 Å, which corresponds to 5.4 Å per turn. This value did not agree with the experimental value, given by the meridianal arcs on the X-ray diffraction patterns, 5.10 Å. I tried to find some way of adjusting the bond lengths or bond angles so as to decrease the calculated distance from 5.4 Å to 5.1 Å, but I was not able to do so.

I was so pleased with the alpha helix that I felt sure that it was an acceptable way of folding polypeptide chains, and that it would show up in the structures of some proteins when it finally became possible to determine them experimentally. I was disturbed, however, by the discrepancy with the experimental value 5.10 Å, and I decided that I should not publish an account of the alpha helix until I understood the reason for the discrepancy. I had been invited to give three lectures on the molecular structure and biological specificity in Cambridge University, and while I was there I talked with Perutz about his experimental electron density distribution functions for the hemoglobin crystal that he had been studying. It seemed to me that I could see in his diagrams evidence for the presence of the alpha helix, but I was troubled so much by the 5.1 Å value that I did not say anything to him about the alpha helix.

On my return to Pasadena in the fall of 1948 I talked with Professor Corey about the alpha helix and the gamma helix, and also with Dr. Herman Branson, who had come for a year as a visiting professor. I asked Dr. Branson to go over my calculations, and in particular to see if he could find any third helical structure. He reported that the calculations were all right, and that he could not find a third structure. More than a year went by, and then a long paper on ways of folding the polypeptide chain, including helical structures, was published by W. Lawrence Bragg, John Kendrew, and Max Perutz, in *Proceedings of the Royal Society of London*. They described about twenty structures, and they reached the conclusion that none of them seemed to be satisfactory for alpha-keratin. Moreover, none of them agreed with my assumptions, in particular the assumption of planarity of the peptide group. (Lord Todd has told the story of his having told Bragg, when they were just beginning their work, that the main-chain carbon-nitrogen bond has some double bond character but that Bragg did not understand that that meant that the peptide group should be planar.)

My efforts during a year and a half to understand the 5.1 Å discrepancy had failed, but Dr. Corey and I decided that we should publish a description of the alpha helix and the gamma helix. It appeared in the *Journal of the American Chemical Society* in the fall of 1950. It was followed in 1951 by a more detailed paper, with Branson as coauthor, and a number of other papers on the folding of polypeptide chains. An important development had been the publication of X-ray photographs of fibers of synthetic polypeptides, in particular of poly-gamma-methyl-L-glutamate, by investigators at Courtaulds. These striking diffraction photographs showed clearly that the pseudo repeat distance along the fiber axis is 5.4 Å rather than 5.1 Å. There are strong reflections near the meridianal line, corresponding to 5.1 Å, but they are not true meridianal reflections. On the X-ray photographs of hair the reflections overlap to produce the arc that seems to be a meridianal reflection. It was this misinterpretation that had misled all of the investigators in this field. It was accordingly clear that the alpha helix is the way in which polypeptide chains are folded in the alpha-keratin proteins.

Moreover, we reached the conclusion, as did Crick, that in the alpha-keratin proteins the alpha helices are twisted together into ropes or cables. This idea essentially completed our understanding of the alpha-keratin diffraction patterns.

The apparent identity distance in the fiber X-ray diagrams of silk is somewhat smaller than corresponds to a completely extended polypeptide chain. We accordingly concluded that the polypeptide chains have a zigzag conformation in silk and the beta-keratin structure. We reported in detail three proposed sheet structures. The first one, which we called the rippled sheet, involves amino-acid residues of two different kinds, one of which cannot be an L-amino-acid residue, but can be a residue of glycine. It was known that Bombyx mori silk fibroin has glycine in 50 percent of its positions, with L-alanine or some other L-amino-acid residue (such as L-serine) in the alternate positions, so that the rippled sheet seemed to be a possibility for Bombyx mori silk fibroin. It turned out, however, that Bombyx mori silk fibroin has the structure of the antiparallel-chain pleated sheet. The third pleated-sheet structure, the parallel-chain pleated sheet, is also an important one.

About 85 percent of the amino-acid residues in myoglobin and hemoglobin are in alpha-helix segments, with the others involved in the turns around the corners. In other globular proteins the alpha helix, the parallel-chain pleated sheet, and the antiparallel-chain pleated sheet all are important structural features. These three ways of folding polypeptide chains have turned out to constitute the most important secondary structures of all proteins. Dr. Corey, to some extent with my inspiration, designed molecular models of several different kinds that were of much use in the later effort to study other methods of folding polypeptide chains. I used these units to make about one hundred different possible structures for folding polypeptide chains. For example, if the hydrogen bonds are made alternately a little shorter and longer than 2.90 Å in a repeated sequence, an additional helical twist is imposed upon the alpha helix. Some of the models that I constructed related to ways of changing the direction of the axis of the alpha helix. I reported on all of this work at a protein conference in Pasadena in 1952, but then I became interested in other investigations and stopped working in this field.

It pleases me to think that our work in the division of chemistry and chemical engineering at Caltech, first in collecting experimental information about the structure of molecules, then in developing structural principles, and finally in applying these principles to discover the alpha helix and the pleated sheets, has shown how important structural chemistry can be in the field of molecular biology.[11]

Elsewhere, in talking to laypersons, Pauling had an easy way of summing up the reason for his inspirational accomplishment as he lay in his sickbed in Oxford in 1948.

WE had nothing in 1948 that I didn't know in 1937. I realized I had all the data I needed back in 1937, and I just hadn't thought it out enough.

So I stopped asking the question I'd been asking all along, which was, "What would a polypeptide chain look like if it satisfied a requirement that is indicated by the X-ray diffraction pattern?" Now I asked the question, "How would I fold if I were a polypeptide chain?" Two hours later I had the answer! It was very satisfying.[12]

A succession of papers by Pauling and colleagues on the structure of proteins began appearing in 1951 in the *Proceedings of the National Academy of Sciences (PNAS)*. Meanwhile, as Pauling concentrated on protein structures, his theories and discoveries were being adapted by a small team of researchers in Cambridge, England.

In a 1983 lecture at UC Berkeley, Pauling briefly noted the sequence in discovery that led to DNA, showing some of his role in the process.

IN 1940 the German physicist Pascal Jordan published a paper in which he said that quantum mechanics required that two identical molecules interact more strongly than two nonidentical molecules, that this interaction explained biological specificity, and that it caused a gene to reproduce itself—that is, a gene causes an identical gene to be formed.

Dr. Delbrück brought this paper to me. He and I then published a paper in which we showed that the special interaction between like molecules is far too small to do the job. We wrote that the gene consists of two mutually complementary strands, which can unfold to permit each to act as a template for the formation of a replica of the other.

Thirteen years later, Watson and Crick discovered the double helix.[13]

Much has been said speculatively by scientists and scholars about Pauling's failure to discover the structure of deoxyribonucleic acid—DNA—since he of all researchers was in the best position to do so. Deciphering DNA would unlock the genetic codes contained within chromosomes and genes, already known to carry the instructions governing both species and cell replication in individual life forms. Unlike James Watson and Francis Crick, however, during the early 1950s Pauling was not operating in a single-minded and highly competitive mode. Some major distractions took him away from pure research (see Part III). However, in early 1953, a few months prior to the publication of Watson and

Crick's paper, Pauling had ventured to suggest a structure for DNA: *triple* helix. He jumped into print prematurely on this—something he had sedulously avoided earlier in his career. The three-stranded helix soon proved wrong. Pauling's team had made miscalculations. The proposed structure also had a second flaw—the assumption that the helical strands were held together by hydrogen bonds between phosphate groups. (Actually, the hydrogen atoms necessary for these bonds are absent because of ionic dissociation from the phosphates.)

Pauling may have felt sorely disappointed, even embarrassed, that others had bested him in the quest for the DNA structure. But he claimed that he actually had no sense of being in a race with anyone to discover it. The best explanation for his failure was that he, unlike Watson and Crick, had not seen Rosalind Franklin's new X-ray diffraction patterns of DNA, which were sharper and clearer than the earlier ones that Pauling had relied on. These patterns could be interpreted by indirect reasoning to suggest that the DNA helix had two strands, not three as Pauling surmised from the earlier data. These new data were available during a conference in London in 1952 that Pauling had not attended (though he was to be a key speaker) because the U.S. State Department repeatedly refused to renew his passport on political grounds; he could not leave the country in that time period (see Chapter 7). Pauling sometimes said that had he viewed the new X-ray evidence, he would certainly have rechecked his reasoning and might well have found the correct helix structure. He had been considering various helical forms of DNA, whereas Watson and Crick concentrated on a single helical form, the double helix.

In any case, Pauling graciously credited the discoverers for the simple elegance of their find. He had not said much at all about proteins and protein structure in previous editions of his basic chemistry textbooks because of the lack of progress in determining how they were put together. But when he updated his *General Chemistry* textbook in the mid-1950s, he presented a whole chapter on biochemistry. It discussed how crucial proteins were to life—above all, in the reproductive powers of DNA molecules within all cells, which carry precise genetic instructions.

Pauling summarized the knowledge of DNA and then discussed its impact on science and medicine in a chapter called "Our Hope for the Future" that he contributed to a book on birth defects published in 1963.

THE astounding recent developments in molecular genetics give great promise for the future so far as congenital defects of genetic origin are concerned. Only a decade ago the discovery of the detailed structure of the molecule of DNA, constituting the gene, was made by James Watson and Francis Crick. The four units of DNA, A (adenine), T (thymine), G (guanine) and C (cytosine), occur in each gene (part of a molecule of DNA) in a linear order, a chain, with sequence characteristics of the gene. This chain is twisted around another chain, with a complementary structure: with T in the second chain attached by hydrogen bonds to A in the first chain, A to T, C to G, and G to C. The double chain contains the genes; it is believed that in the process of cell division the chains separate, each serving as a template for the other. Moreover, the sequence of units A, T, G and C provides the information that permits the synthesis of the polypeptide chains of proteins, each with a corresponding sequence of amino acid residues. Rapid progress is now being made in the elucidation of the code, the fundamental code of life—the groups of units of the DNA corresponding to the respective amino acid residues.

It is impossible to imagine the progress that might result from increased knowledge about the structure of DNA and the nature of the molecular mechanisms involved in heredity. This field of science is an extremely difficult one, in part because of its complexity—a single DNA molecule may contain several hundred thousand atoms, and the set of genes of a human being may consist of as many as one hundred thousand of these molecules, each important and each different from all others, except possibly the corresponding molecule in the companion chromosome. But the broad problem of molecular genetics now can be attacked with prospect of continued success. All that is needed is support —the laboratories and the salaries for the scientists. The attack on this problem should continue for decades, for generations. We must keep in mind even the distant possibility that the abnormal DNA molecule of a genetically defective child might be replaced by its normal counterpart or by a surrogate DNA molecule to prevent the manifestations of the congenital defect.

Other possibilities, perhaps not so far in the future, are the introduction of normal cells to replace some of the defective cells, and the manufacture and use of artificial enzymes to replace the enzymes that are lacking because of the gene defect.

The DNA molecules have extraordinary power. The nature of every human being now living has been largely determined by his initial complement of perhaps one hundred thousand DNA molecules, which contain an unknown number of genes in the twenty-three chromosomes of the sperm and the twenty-three chromosomes of the ovum, combined together in the fertilized ovum, which duplicate themselves over and over again as the process of cell division occurs to result in the human organism. These one hundred thousand molecules, although they are large molecules, as molecules go, together constitute a very small amount of matter; if duplicates of the one hundred thousand DNA molecules that have determined the nature of every person on earth, of the 3 billion members of the human race today, could be brought together, the total amount of DNA, constituting the pool of human germ plasm, would be about the size of a pinhead. A similar amount of DNA will determine the nature of the next generation of human beings.

I marvel at the fidelity with which these molecules duplicate themselves. Of course, they have achieved their virtuosity through a long process of trial and error, over the period of about 1,000 million years since the origin of life on earth. It is likely that the first self-duplicating molecules did a poor job, so that many of the molecules of the second generation differed from those of the first. We can see that a process of selection might well occur that would lead to the survival of those molecules that excel in self-duplication. It is estimated by geneticists that on the average, in human beings, a gene undergoes a mutation about once every one hundred thousand generations. For example, the dominant mutant gene that causes achondroplasia (dwarfism) occurs in about one birth in twelve thousand, corresponding to a mutation rate of one in twenty-four thousand (either of the two genes involved may mutate). For other genes somewhat lower mutation rates are usually reported.

Moreover, the amount of abnormality of the mutated gene is small, in the molecular sense. For example, we may say that the gene defect that is responsible for sickle-cell anemia amounts to an abnormality of less than 1 percent of the DNA molecule. The hemoglobin molecule contains two polypeptide chains of one kind, the alpha chains, that are synthesized by one gene, and two chains of another kind, the beta chains, synthesized by a second gene (each of the genes present in double dose). Sickle-cell anemia is caused by an abnormality in the gene that synthesizes the beta chain. The beta chain contains 146 amino acid units, and only one of the 146 units in the beta chain of sickle-cell anemia is different from the beta chain of normal hemoglobin. It is accordingly likely that the gene responsible for the manufacture of the abnormal beta chain is itself abnormal to the extent of only one part in 146. These two genes constitute only 1/50,000 of the genetic material, and, accordingly, the child with sickle-cell anemia is defective in his genetic constitution to the extent of only 1/50,000 of 1/146; that is, his genetic abnormality amounts to only about 1/10,000,000 of the genetic material that he has inherited.

From knowledge of the nature of the forces operating between molecules, I find it astounding that the processes of reduplication of the genes and the synthesis of protein molecules involve so few errors. Can we hope that these processes can be improved, that there will be a decrease in the number of defective children born in future generations? Can we take actions that will increase the yield of human beings free from serious or significant congenital defects?

I believe that man is now changing the nature of the world in such a way as both to decrease and increase the number of defective children.[14]

At this time Pauling often expressed deep concern over the probability of increased defects among children.

WE have been living for some decades in an environment that contains ever-increasing amounts and varieties of molecules that were not present during the period of evolution of the human race. These are the molecules of substances that do not occur in nature and have been only recently synthesized by chemists for use as insecticides, as additives to foods, as additives to gasoline, as drugs, and for other purposes. Some of these substances are toxic to human beings.

He also attributed more genetic defects to radiation from various sources, including his current and abiding misgivings over nuclear contamination.

In the last part of the UC Berkeley Hitchcock Lecture, "Chemical Bonds in Biology," that Pauling gave in 1983, he made his move from scientific matters to societal concerns, which by then was fully expected of him.

I had been pretty well satisfied in 1940 with the idea that complementariness in detailed molecular structure is the secret of life. Now I was nearly completely satisfied.

If I were giving another lecture I could go on in detail. We can imagine how on the primitive earth there developed the "hot thin soup" of Oparin and Haldane, as ultraviolet light and lightning caused almost every moderately simple molecule to form. With the situation far from equilibrium, a catalyst could preferentially accelerate some chemical reactions. Adenine might speed up the production of thymine, and thymine that of adenine. Polypeptides and polynucleotides would come into existence, then enzymes and single-celled organisms. Over a period of a couple of billion years there would be keen competition of the single-celled organisms, with the development of millions of different enzymes, and now we have reached the present stage.

I am happy to feel that I now have an understanding of at least the basic answers to the question "What is life?"

Biochemists and other scientists here in this great university are working on many details of the astoundingly complex system of living organisms here on earth. I take pleasure in learning about their discoveries, and especially in recognizing that all of the phenomena that they discover are based on the complementariness in structure of the interacting molecules.

Now a remark about the double helix.

My wife once said to me, "If that was such an important problem, why didn't you work harder at it?"

I shan't try to answer the question. It just leads up to another statement she made to me, which has in fact determined much of my activity.

She said, "The most important of all activities is that of keeping the world from being destroyed in a nuclear war."

And so I conclude by saying to you: "The most important of all problems is that of keeping the world from being destroyed in a nuclear war." Every one of you should do whatever you can to prevent this ultimate catastrophe.[15]

From the end of World War II on, Linus Pauling devoted a good portion of his time, energy, and intellect to addressing the prospect of nuclear war that now faced humanity, possibly to bring on its own extinction after an evolutionary path that had taken several billion years. Pauling's activities launched a radically new research and development program in his life that required social activism and another means of attack in solving problems. In the process the scientist became a notable humanist.

The greatest worldly reward for scientific discovery, in both international prestige and professional recognition, is the Nobel Prize, which in the sciences is given for Chemistry, Physics, and Physiology or Medicine. (The prizes in Literature, Peace, and, more recently, Economics are of course the equivalent humanistic awards.)

Pauling through time had made a number of contributions and discoveries in scientific research areas that spanned several Nobel categories. Colleagues kept expecting him to receive the award, but it proved elusive. Pauling himself said that he didn't really expect it because the prize was given for a notable discovery, not for cumulative discoveries or a considerable body of work. Scientists often say that Pauling might have received the Nobel in Chemistry for any number of earlier discoveries. (It is also declared that Pauling well deserved the Nobel Prize in Medicine or Physiology for his discovery of the molecular cause of sickle-cell anemia, which launched biomedical probes into innumerable molecular diseases.)

In any case, Pauling was finally given the Nobel Prize in Chemistry in 1954, the year after Watson and Crick announced their discovery of the structure of DNA. Pauling's award seemed almost like a consolation prize for his failing to determine DNA's structure: after all, his protein work had laid the foundation. Actually, Watson and Crick had to wait until 1962 for their Nobel honors; by then the validity of their proposed structure had been well verified.

Pauling's Nobel speech, "Modern Structural Chemistry," which he gave in Stockholm in 1954 when he received the illustrious gold medal, mostly recapitulated previous lectures and articles reviewing the technical history of research on chemical structure and bonding. But his final paragraph appropriately addressed the future.

WE may, I believe, anticipate that the chemist of the future who is interested in the structure of proteins, nucleic acids, polysaccharides, and other complex substances with high molecular weight will come to rely upon a new structural chemistry, involving precise geometrical relationships among the atoms and the molecules and the rigorous application of the new structural principles, and that great progress will be made, through this technique, in the attack, by chemical methods, on the problems of biology and medicine.[16]

Pauling's Nobel Prize in Chemistry gave him a certain fame, prestige, and authority that he had lacked up till now beyond the scientific community. He would employ this new prominence mostly in his antinuclear and peace-promoting activities.

III

The Nuclear Age

1945–1994

Seven

Atomic Politics

The release of atomic energy has changed the nature of the world, and the mechanisms of international relations will have to be changed accordingly.

Many people who admired Linus Pauling have thought that his intense interest in world peace began in the period following World War II, when he was concerned about atomic weaponry and saw the need for developing a means to resolve international conflicts. His detractors believed he would willingly capitulate to a foreign power notorious for imposing its economic, social, and belief systems on other nations—and in fact that he had become a turncoat to his own country. Certainly none of them must have heard his rousing yet well-reasoned talks in the late 1930s and the early 1940s, when he urged Americans to save democracy from collapse in Europe by joining with England and the British Commonwealth nations to fight against the Nazi takeover.

Before the laggardly United States finally entered World War II, Pauling gave public speeches to educate people about the need for democracies to band together to combat the threat of Hitler's Germany ruling the world. He handwrote this speech he gave on July 22, 1940, at a Pasadena gathering. Its words convey his alarm over the collapse of the European democracies, with Great Britain left as a besieged, holdout nation.

THE goal of our social development is freedom—freedom of individual action, freedom of speech, religious freedom, and freedom from the fear of unjust and arbitrary oppression and persecution. By centuries of struggle and sacrifice this goal has been in its essential achieved in the democracies of the world.

Now there is being waged a great war between democracy and totalitarianism, to decide between the free way and the slave way. And this war may well determine, as Hitler says it will, the course of the world for the next thousand years. Through the development of methods of transportation and of technology in general the world has effectively become so small that world rule is to be expected soon. The great decision which will be made before many years—surely during the present century, and possibly within the coming decade—is whether the world will be ruled by totalitarian masters or whether it will be a free democratic state.

We have become accustomed to this freedom, and we had begun to accept it without question, as something that was given us for nothing, something for which we do not need to struggle and fight. But events are showing us that this is not right—that the fight for freedom is not yet over. During the last few years we have seen many of the people of Germany persecuted and robbed of their freedom and their lives, and in recent months it has been the people of peaceful democracies who have been enslaved and killed.

In a democracy, such as ours, there is little danger of loss of freedom from actions within the country. The democratic system is, to be sure, slow and unwieldy and inefficient, but it is safe, in consequence of the fact that the important decisions are made by the people themselves. The one great danger that threatens a democratic country is the danger of attack and overthrow by militant enemies with the lust to conquer and enslave. . . .

There is the chance, which is perhaps a large chance, that Britain, despite the desperate resistance which she will make, will be overthrown, and that her fleet will fall into the enemy's hands. Then Hitler could use the British fleet and merchant marine to bring the war to America; and he might well decide to attack his one remaining enemy, the United States, at once, to catch us while we are still only very poorly prepared, rather than wait until we have built up the great military machine for which our plans are now made. . . .

This is a war whose outcome involves the decision between freedom and slavery for the people of the world; it is a struggle between democracy and the enemies of democracy. We all recognize that this is our war, and that Britain is fighting our fight—just as did Norway and Holland and Belgium and France until they were overcome. Recognizing this, the President and the great majority of the people favor the policy of the government of advancing all possible material aid to Britain. . . .

There is another thing that we must consider. This step would mean going to war, and we, as idealists, are by nature pacifists and opposed to war. But we are being forced into war anyway—we are vigorously preparing for war, and who among us believes sincerely that we are not going to have war sooner or later? We as individuals do not believe either in fighting with our fists or clubs. But we would fight a thug who attacked us. And if the thug were to attack a peaceful neighbor of ours, would we not come to his aid rather than wait until he had beaten our neighbor into insensibility and had begun on us? This is the situation of the United States. Should not our country help Britain now to fight off the thug who is attacking her and will most probably attack us when she is polished off? . . .

When we consider that during the last few months half the democracies of the world have been brutally and ruthlessly conquered or surrounded and enslaved by the antidemocratic dictatorial totalitarian forces of evil, that the British democracies are now fighting off the aggressors and may be overpowered and conquered during the coming months, despite the aid which we are giving them, that we in the United States are beginning a desperate attempt to prepare for the war that we see ahead of us, must we not admit that it would be the part of wisdom for us to recognize our unity of purpose with democracy throughout the world, to form at once a federal union with the remaining free democracies, and thus to make the united stand which would win for the side of freedom the great battle for the world of the future?[1]

This was not a man who would betray the principle of democracy or relinquish freedom of thought and the expression of belief—his own or anybody else's—or would compromise fact and truth to accommodate some dogma. Moreover, Pauling's experiences during the Second World War in conducting research projects for the government and in participating on science committees added another major component to his life: a strong commitment to public service. From his college years on, he had felt that everybody—scientists not excepted—had obligations to society. Now he was discovering important ways to be useful.

Pauling was particularly proud of his wartime work for the Office of Scientific Research and Development (OSRD), whose director, Vannevar Bush, had been asked by President Roosevelt in 1944 to produce a comprehensive plan for the future of American science and medicine. As history shows, wars tend to greatly accelerate technological advances, and Roosevelt, impressed with the scientists' roles in the war effort, wanted to maintain the impressive momentum. "The information, the techniques, and the research experience developed by the Office of Scientific Research and Development and by the thousands of scientists in the universities and in private industry," he wrote, "should be used in the days of peace ahead for the improvement of the national health, the creation of new enterprises bringing new jobs, and the betterment of the national standard of living."

In response, Bush created special panels of experts to investigate different scientific issues. Pauling accepted an appointment to the Medical Advisory Committee and was the only nonphysician among the eight appointees. He had an interesting time traveling around the nation with them, visiting medical schools, hospitals, and biomedical research centers to talk with medical people. The experience gave him better direct acquaintance with health problems of all kinds.

The Bush report, *Science: The Endless Frontier,* was delivered to President Truman in 1945. It made numerous recommendations. Not long afterward Pauling commented: "As a member of the Medical Advisory Committee, I had an opportunity to see how great an effort has been made to find a way of providing the needed federal aid to research in science and medicine without invoking the evils such as political influence and mediocrity of performance which may characterize government activities."[2]

World War II, however, had created a training and manpower problem in the sciences that Pauling noted in 1945 in an issue of Caltech's monthly magazine, *Engineering and Science.*

THE great contributions to the war effort made by the scientists during the past five years have made the public aware that the welfare of the nation depends upon adequate support of research in science and medicine. Moreover, we are beginning the postwar period with a great deficit —not only a deficit of trained scientific personnel, resulting from the interruption of the education of the 165,000 men who, except for the war, would have received scientific or engineering degrees during the war years, and were prevented from receiving those degrees, but also a deficit in the body of fundamental scientific knowledge. The contributions made by the scientists during the war years were very largely based on fundamental discoveries made before the war. Further progress in industrial development and in medicine will be possible only if proper support is provided for basic scientific research.

It has been clear that the sources of funds drawn upon in the past for support of basic scientific research will not be adequate in the immediate future.[3]

When Pauling became president of the American Chemical Society in 1949, his first message to members in the ACS magazine, *Chemical and Engineering News,* was about balancing the association's budget, which could "without doubt be met in a straightforward way." The next agenda item he presented was far more complex.

BUT the greatest problem of all is the one that has resulted from the past inability of man and his social and political organizations to give rise to the challenge of nature by discovering the way for all men to live together in peace and to use the riches of the world for the general welfare and happiness. The present grievous state of poverty over most of the world is the result in part of the destruction and waste of World War II and of the use of a seriously large fraction of our natural resources and our capacity for production in the attempt to meet the threat of another cataclysmic conflict.

How can the American Chemical Society help in the attack on this great problem? I believe that it can help in many ways: by continuing to foster increased understanding and friendship among chemists of different nations [and] by cooperating with UNESCO in the support of international fellows and international scientific meetings.[4]

Then Pauling cited a crisis in the postwar funding of scientific research and came up with some strong proposals.

EVERYONE now knows that the practical contributions of science depend upon fundamental scientific discoveries, and that the rate of practical progress will fall off if basic scientific research is not continued. And yet at the present time an inadequate amount of support is being provided for research in pure chemistry and in other fields of pure science —our wellsprings of fundamental knowledge, instead of being treated to cause them to give the greatly increased flow that we can see is needed for the thirsty modern world, are in danger of drying up. The colleges and universities, in which most pure research is carried out, have in effect, because of the decrease in value of the dollar, a decreased return from endowments for support of research, and the contributions of private foundations have also effectively decreased in amount. It has been evident since the publication of the Bush report that we must have a national science foundation, and that it must have as its minimum basis of operation the scale recommended in the report, amounting (with a correction for the increase in costs) to an annual expenditure of at least $100 million the first year (the fifth postwar year), and increasing in about two years to $250 million. By allocating some of their funds to the support of research in pure science, the armed services and the Public Health Service have helped to meet the crisis caused by the failure of the government during the past three and a half years to set up the foundation, but this is a temporary measure. We shall have failed if an effective national science foundation is not created and put into operation during the coming year.

I believe that, since all the people benefit ultimately from scientific discoveries, it is proper that a large amount of support of fundamental research be provided by the federal government. There is danger, however, in having the support of basic research come exclusively or predominantly from a single source. The colleges, universities, and pure research institutes should receive significant support for research also from the state governments, from permanent endowments, from the great private foundations, and from industry.

The industrial corporations, the chemical industries that depend upon fundamental science for their success, are, I believe, failing to do their part in the support of pure chemistry. I agree with many leading industrial scientists in thinking that it would be proper for these corporations, instead of having university chemists help with the solution of special technical problems, to support not 1 percent, but 10 or 20 or 30 percent of the research in pure chemistry carried out in the universities of the country, by making research grants without any restriction on the nature of the research and without patent restrictions or restrictions on the publication of results and dissemination of information, and by making these grants to the poorer and at present less effective schools as well as to the richer and better ones. The year 1949 would be a year of great progress indeed if it were to witness not only the formation of an effective national science foundation but also the adoption of a sound and significant program of industrial support of fundamental research in our colleges and universities.[5]

Pauling spoke and wrote enthusiastically elsewhere about these proposals, most notably in another ACS presidential address, "Chemistry and the World of Today," in which he stirred up a storm of protests by publicly supporting socialized medicine.[6] Soon afterward, in 1950, the National Science Foundation was created, to generate and fund projects at independent institutions. The National Institute of Health (NIH) became massive and plural (Institutes) as it began expanding in the 1950s, subdividing into a dozen institutes, each with its own areas of specialization—all conducting research within their own premises but also subsidizing other biomedical research across the nation.

· · ·

In the late summer and autumn of 1945, the first few months after the United States dropped two atomic bombs on Japan to bring the war to a swift and dramatic end, Pauling gave talks to various citizens' groups. Because of his own theoretical work in structural chemistry, Pauling had sufficient knowledge of basic nuclear physics to help educate people about this newly harnessed atomic power—its potential for useful energy and other positive gains, but above all its fearful use in weaponry, brought to deadly fruition in the code-named Manhattan Project. In the first matter-of-fact public lectures that Pauling provided, he apparently refrained from making value judgments or prognosticating on the likely course of warfare or international relations in the new atomic era.

THEIR significance as a source of energy has also been pointed out, by the statement that one pound of uranium or thorium is equivalent to two and a half million pounds of coal. When we remember that uranium and thorium are not rare elements, but are among the more common elements—the amount of uranium and thorium in the earth's crust being about the same as that of the common element lead—we begin to understand the promise of nuclear energy for the world of the future, and the possibility of its contributions to the well-being of man. I believe that it will soon be recognized that the discovery of the controlled fission of atomic nuclei and controlled release of atomic energy is the greatest discovery that has been made since the controlled use of fire was discovered by primitive man.[7]

Moreover, Pauling was noticing that the very subject of atom bombs created a breakthrough in public interest in rudimentary nuclear physics, even a craze. To become familiar with popular thinking about science so he could be au courant in his lectures to both scientists and laypersons, he studied news articles. In his acceptance speech to fellow chemists when he was awarded the ACS's Theodore William Richards Medal in Chemistry in 1947, Pauling launched his discourse "The Unsolved Problems of Structural Chemistry" with some levity.

IN our present atomic age everyone is continually made aware of the existence of atoms and molecules. Atomic fission and other atomic phenomena are presented to the younger generation in the comic papers, and molecules and their properties are introduced to us in advertisements. The schoolboy now accepts the existence of atoms and molecules without question, and he is apt to have a reasonably good understanding of their properties and significance before undertaking the study of science in school.

I have noticed that the advertisements in our national magazines sometimes present problems, problems that have remained unsolved. One of the problems which has interested me is the nature of "pinpoint carbonation." I have been given the impression that this phenomenon is interesting and valuable, but I have not been able to discover its nature.

Another phenomenon of which I have tried, without success, to obtain an understanding is the "activation" of chlorophyll in the household deodorizer Airwick. The selection of chlorophyll to be used in the preparation is no doubt to be attributed to its well-known ability to purify the air by converting carbon dioxide to oxygen. I have been interested in the properties of chlorophyll for some time, and I had not heard from other sources about the activation of chlorophyll. By obtaining a copy of a patent I learned that chlorophyll is activated by formaldehyde. This type [of activation] seems to differ somewhat from ordinary types, because a great many molecules of formaldehyde are required for the activation of each molecule of chlorophyll. The thought has occurred to me that the significant phenomenon may really be one of deactivation, rather than activation, and that the formaldehyde may be acting in a way similar to that effective when it is used in embalming fluid.[8]

Growing serious, Pauling gave a detailed consideration of many areas into which researchers in structural chemistry might probe, and then reached his grand conclusion.

THE progress of science in recent years is bringing biology and medicine into closer and closer contact with the basic sciences, and I am confident that the next few decades will bring to us a detailed understanding of the molecular structure of biological systems, and that this understanding will help in the rapid general progress of biology and medicine.[9]

The discovery of DNA, which would greatly accelerate the advance of biomedicine and biology, lay only six years away. But already Pauling's attention was being pulled away from laboratory research. He had started his all-out campaign for international peace and against any possibility of nuclear warfare. He played all three roles at once—scientist, educator, and reformer.

Almost fifty years later Pauling reflected on this early postwar period.

I did not have any secret information. I was free to talk, or at least I thought I was. When a group of businessmen asked me to give a talk explaining what nuclear fission is and what the Hiroshima and Nagasaki atomic weapons were, I accepted the invitation and gave the talk. The next day there was an FBI man in my office, who said, "Where did you get the information as to how much plutonium there was in the [Nagasaki] atomic bomb?" I replied that I had figured it out, so he went away and did not come back. At least *he* did not come back, but the FBI came back of course in another form![10]

In his information-dispensing activity, however, Pauling was arousing the suspicion of government agents. He might, after all, be revealing secrets useful to alien interests. Here is what happened when he gave the talk "Atomic Energy and World Government" at the Hollywood-Roosevelt Hotel on November 30, 1945. By this time Pauling clearly was concerned with the implications of the atomic age to international relations, a subject already of keen interest to him before the war. He had also become involved with the new Federation of Atomic Scientists, which on November 6 had issued a forceful "Statement on Atomic Power" that expressed the need for an international system of control and cooperation to deal with the fact that "there has been unleashed in the atomic bomb the most destructive force known to mankind, against which there can be no military defense and in the production of which the United States can have no enduring monopoly." The carbon copy of Pauling's speech contains handwritten brackets that were inserted afterward around sensitive information that a government censor, an intelligence officer with the Manhattan Project, warned Pauling to delete in any future talks he gave. (These are duplicated below.)

THE optimist thinks that this is the best of all possible worlds—the pessimist fears that this may be so.

It is the atomic bomb which is responsible for my being here tonight —just as the atomic bomb may be responsible for our all not being here a few years from now. Like most other scientists, I have in the past stuck pretty close to my work and paid little attention to politics and to world affairs, perhaps even less than the average citizen. Most scientists have a deep interest in their work—in the job of discovering the nature of the physical world—and they have little time for or interest in such crude activities as politics and world affairs. Now, however, scientists have made a discovery which truly revolutionizes the world. No one understands how great is the significance of atomic energy and the atomic bomb to the world so well as do the scientists, and this understanding has forced them into activity, has caused them, individually and in groups such as the Federation of Atomic Scientists, to begin a campaign of talking, of presenting the facts about the atomic bomb, in the hope that everyone, as he begins to understand the possibilities of the future, will be horrified by them, and will pledge himself to take the individual action which is necessary to save the world.

I have never seen an atomic bomb explode, but several of my friends have, and have told me about it. The limited experience which I have had during the war with ordinary explosives has not been of much help to me in appreciating the power of atomic explosives, because ordinary explosives are hardly worth mentioning in the same breath with fissionable atoms. And yet we used to think of nitroglycerine and TNT and RDX and PETN as pretty powerful substances, worthy of respect: a pound of TNT, in the form of a shaped charge, can blow a hole through six inches of armor plate; a hundred pounds, which might be carried into this room by one man, could kill everybody in the room; a few tons, which might be dropped as a bomb, could destroy this building or devastate a whole block of houses.

TNT and similar molecular explosives are powerful because the atoms in their molecules are combined in such a way that a large amount of energy is stored up, which can be suddenly released as a detonation wave causes the atoms in the molecules to rearrange themselves into new product molecules, which fly apart with great velocity, impinging on surrounding molecules and producing the shattering shock wave which spreads out from the center of the explosion.

The energy of TNT is stored up between the atoms in the molecule, and not within the atoms. Only six years ago was it discovered that there is stored up in the minute nuclei of heavy atoms themselves an almost unbelievably great amount of energy which can be released at the will of man. These heavy nuclei are themselves unstable, in the same way that molecules of TNT are unstable; under the influence of neutrons these nuclei can split in two, with the evolution of an incredible amount of energy, which causes the split products to fly apart with terrific velocity.

It is in the nature of this reaction, its dependence upon a supply of neutrons, that it proceeds slowly if there are not too many fissionable atoms around—uranium 235 or plutonium—but that it proceeds explosively, in a millionth of a second, whenever more than a certain amount, [a pound or two,] of the material is brought together. Suppose that we had two pieces of plutonium, each shaped like half of a golf ball [and weighing perhaps a pound (the density of plutonium is very great— about twenty times that of water),] and a mechanism for suddenly clapping them together: this would be an atomic bomb, like that which devastated Nagasaki, and if it were to explode here, it would destroy Hollywood.

This, then, is an atomic bomb—the smallest that we know how to make now—[a couple of pounds of plutonium,] equal to *forty million pounds of TNT*, and capable of wiping out a city ten square miles in area and killing hundreds of thousands of people. Yesterday Harold Urey testified that three such bombs exploded in Washington could cause our federal government to vanish. And *these* are *little* atomic bombs, the smallest that will explode—a big one, which the atomic bomb scientists say could be made, could flatten the whole of New York and kill its millions of inhabitants.

[We must not think that because only three atomic bombs have been exploded there are only a few in existence. I can make a guess about this, on the basis of the order-of-magnitude figures quoted in the Smyth report—and my guess is that there are perhaps a hundred or two in existence—enough to kill perhaps fifty million people, if each were dropped on a different city—and by next year there may be five hundred.] My colleague Robert Oppenheimer, whose opinion is based on knowledge of the situation, has stated that he expects tens of thousands possibly to be used if there is an atomic war.

Nor must we think that atomic warfare will be expensive. The atomic bomb program cost two billion dollars, and only two bombs were dropped on Japan—but Oppenheimer has said that in mass production these bombs would cost not a billion dollars apiece, but only a million dollars. This is terribly cheap—a million-dollar atomic bomb has the power of forty million pounds of TNT, which in bombs would cost perhaps $100 million dollars. For a given amount of money we can do a hundred times as much damage to the enemy as in preatomic days. The next war should be a real one, indeed; perhaps it will be equal to one hundred like the one we have just gone through.

Offensive action will be easy and cheap, but defensive action will be hard and expensive. No good means of defense exists, and it seems inconceivable that an effective one will be found.

Ladies and gentlemen, this is the danger now facing the world. What can be done to avoid it? Scientists discovered the basis of the atomic bomb and the atomic scientists on the bomb project have had a longer time to think about the problem it poses than other people have had. They believe, and I believe, that there is only one way to avoid world disaster—and that is to abolish war, to have effective international control of the atomic bomb, and, as soon as possible, to form a world government to which the nations of the world give up their sovereignty in matters which serve as causes of war. The Federation of Atomic Scientists has formulated a statement to this end, urging "that the President of the United States immediately invite the governments of Great Britain and the Soviet Union to a conference to prevent a competitive armaments race, to plan international control of mankind's most devastating weapons, and jointly to initiate international machinery to make available to all peoples the peacetime benefits of atomic energy." This statement has been signed by hundreds of workers at Oak Ridge, Hanford, and Los Alamos, and by many hundreds of other scientists in southern California and elsewhere on the Pacific Coast and throughout the nation.

I believe that the only hope for the world is to prevent war, and that war can be prevented only by a sovereign world government to which individuals are directly responsible. Talk about a world government has always been considered visionary—even now there seems to be a feeling, especially among editorial writers, that the practical people of the United States will never consent to this nation's giving up its sovereignty. But is not the formation of a world government, abolishing war, the practical thing to do? Would it not be more realistic, more practical for the United States and other nations to give up some of their sovereignty, that relating to waging war, to a world government than for these nations to retain all their sovereignty and to be destroyed? Is it not more realistic, more practical to use the gifts of nature, as discovered by science, for the good of all the people of the world, considering them as brothers, than for death and destruction? I believe that the discovery of atomic power will be recognized as necessitating world unity, and that the goal of a continually peaceful and happy world, which a few years ago was hardly visible in the greatly distant future, will be achieved within our generation.[11]

Within a few months after his first talks to the public, Pauling said later, "I began introducing the idea of rejecting the very possibility of nuclear war because of the weapons' destructiveness." He changed the very nature of his talks—including the intended effect upon his listeners. He always traced this transformation to his wife's influence. "What good will science do if the world is destroyed, Linus?" she asked him, and he could make no reassuring response.

Pauling wrote about Ava Helen's important part in arousing his social activism in an article entitled "An Episode That Changed My Life," which he submitted to *Reader's Digest* in the mid-1980s. It was never published.

DURING the Second World War I continued my teaching, but also was engaged in many investigations of scientific and medical problems relating to the war effort, including work on explosives. I had been asked by Robert Oppenheimer to join him in the work on the atomic bomb at Los Alamos, but had decided not to do that, and instead to continue the work that I was doing at the California Institute of Technology and as a member of war-research committees in Washington, D.C. In August 1945 atomic bombs were exploded by the United States over Hiroshima and Nagasaki, Japan. Each of these bombs, involving only a few pounds of nuclear explosive, had explosive power equal to fifteen thousand or twenty thousand tons of TNT. The nuclear explosive, plutonium or uranium 235, has 20 million times the explosive power of TNT or dynamite. These two small bombs destroyed the cities and killed about 250,000 people.

Someone who knew that I was an effective lecturer about chemical subjects invited me to speak at a luncheon before the members of the Rotary Club in Hollywood, to tell them about the nature of this tremendously powerful new explosive, involving fission of the nuclei of the atoms. I did not have any classified information about the atomic bombs, and so I was free to speak. My presentation was essentially a scientific one, about the structure of atomic nuclei and the nature of the process of nuclear fission, and also about the Einstein relation between mass and energy, which explains why the splitting of atomic nuclei can result in the release of a tremendous amount of energy—far, far greater than can be released by any chemical reaction, as in the detonation of TNT.

Later I gave a similar talk before another group, in which I discussed not only the nature of nuclear fission but also the change that had occurred in the nature of war through the development of atomic bombs. I quoted Albert Einstein, who had said that the existence of these bombs, so powerful that a single bomb, lobbed over by a rocket, could destroy a whole city, required that we give up war as a means for settling disputes between the great nations, and instead develop a system of world law to settle these disputes. I also quoted statements by various politicians and students of international relations.

After this lecture, when my wife and I had come home, she made the following statement to me:

I think that you should stop giving lectures about atomic bombs, war, and peace. When you talk about a scientific subject you speak very effectively and convincingly. It is evident that you are a master of the subject that you are talking about. But when you talk about the nature of war and the need for peace, you are not convincing, because you give the audience the impression that you are not sure about what you are saying and that you are relying on other authorities.

These sentences changed my life. I thought, "What shall I do? I am convinced that scientists should speak to their fellow human beings not only about science, but also about atomic bombs, the nature of war, the need to change international relations, the need to achieve peace in the world. But my wife says that I should not give talks of this sort because I am not able to speak authoritatively. Either I should stop, or I should learn to speak authoritatively."

I had by this time begun to feel so strongly about these matters that I decided that I would devote half of my time, over a period that has turned out to be nearly four decades, to learning about international relations, international law, treaties, histories, the peace movement, and other subjects relating to the whole question of how to abolish war from the world and to achieve the goal of a peaceful world, in which the resources of the world are used for the benefit of human beings, and not for preparation for death and destruction. During the next years I gave hundreds of lectures about nuclear weapons, the need for world peace, and, from 1957 on for several years, the damage to the pool of human germ plasm and to the health of living people by the radioactive fallout from the atmospheric testing of atomic bombs. My life, ever since that day nearly forty years ago, no longer involved my wholehearted efforts in teaching science and carrying on scientific research. Instead, half of my energy was devoted to that work, and the other half to working for world peace.[12]

Another factor besides his own and Ava Helen's concerns propelled Pauling out into the limelight surrounding the growing debate over the development of nuclear weaponry—the world-famous physicist Albert Einstein. Pauling had first met Einstein in 1927, when the latter attended a Pauling lecture at Caltech—afterward declaring that he would have to learn more about chemistry in order to understand what Pauling had said. ("Linus Pauling—now *there's* a genius!" he is reported to have commented in later years to an acquaintance.) When Einstein returned in 1932 for an extended visit to southern California, during which Caltech's administrators and trustees tried to lure him to join the faculty, Pauling became better acquainted with him. In the postwar period he would get to know Einstein far better because of their mutual concern

over nuclear warfare. Einstein heard about Linus Pauling's lecturing activities, for in 1946 Pauling was asked to join him and six other prominent American scientists on the Emergency Committee of Atomic Scientists. Einstein was the chairman. This is how Pauling at times recollected the experience.

WHEN atomic bombs were first exploded at Alamogordo, and then at Hiroshima and Nagasaki, Albert Einstein was horrified. He had known that something was going on, but did not know exactly what. Perhaps he had not realized that these atomic bombs would have about fifteen thousand times the destructive power of the greatest TNT bombs used in the Second World War. Einstein said, "Now that one bomb can destroy a whole city, we must give up war as a way of settling disputes between nations." He went on to say something that is pertinent now. "We cannot leave this problem to the generals, politicians, and diplomats to work out a solution over a period of generations. It is up to the people to see to it that this nonsense is stopped."[13]

He allowed the vice-chairman to run the [Emergency Committee] meetings and did not attend them himself. But he always requested my wife and me to visit and talk with him. We were the only ones. During our stay at Princeton, we discussed, not science, but mainly world affairs for about an hour every evening. We came to know him quite well. I think he liked my wife especially. They both had excellent senses of humor. . . .

To an extent, he worried about his part in developing the atomic bomb. His theory of the relation between mass and energy was, of course, the basis for the bomb. His famous equation $E = mc^2$ (energy equals mass times the square of the speed of light) showed scientists that, if they could accelerate nuclear reactions, the atomic bomb would be feasible. . . .

Once in a discussion [Einstein] said he felt he might have made a mistake in signing a letter to President Roosevelt recommending production of atomic bombs. He went on to say that perhaps he could be excused because he and his colleagues feared that Germans were working on the problem and would become masters of the world if they came up with a bomb first.

Immediately after saying good-bye to Einstein on that day, I took out my diary and wrote down precisely what he had said to me and the way he had said it.

Perhaps my own work for world peace would not have been very effective if I had not been invited to become a member of the board of trustees of the Emergency Committee of Atomic Scientists. It raised money and made a film showing the Hiroshima and Nagasaki explosions and what damage they did. Leo Szilard and I went around showing the film and talking about what was going on, and saying that we must avert a nuclear war.

Before then I had made some public talks about nuclear weapons and nuclear war; but it was Einstein's example that inspired my wife and me to devote energy and effort to pacifist activities. . . . [We] talked with Albert Einstein many times, and it is my opinion that he thought about the world much the way that I did then and still do.[14]

"The End of War," the first chapter of Pauling's book *No More War!* published in 1958, expressed some of the material that he customarily presented, along with his own feelings regarding the subject. Here was his credo, expressing faith in human intelligence and reasoning yet realistic awareness of the fierce irrationality in human feelings that could ultimately annihilate humankind, now that scientists had created the means for doing so.

IN the past, disputes between groups of human beings have often been settled by war. At first the wars were fought with stones and clubs as weapons, then with spears and swords, and then with bows and arrows. During the last few hundred years they have been fought with guns, and recently with great bombs dropped from airplanes—blockbusters containing one ton or even ten tons of TNT.

There may in the past have been times when war was a cruel but effective application of the democratic process, when force was on the side of justice. Wars fought with simple weapons were often won by the side with the greater number of warriors.

Now war is different. A great mass of people without nuclear weapons, without airplanes, without ballistic missiles cannot fight successfully against a small group controlling these modern means of waging war.

The American people were successful in their revolt against Great Britain because modern weapons had not yet been developed when the American Revolution broke out.

It is hard for anybody to understand how great a change has taken place in the nature of the world during the past century or two, and especially during the last fifty years. . . .

Never again will there be the world of William Shakespeare, the world of Benjamin Franklin, the world of Queen Victoria, the world of Woodrow Wilson.

The scientific discoveries that have changed the world are manifold. I think that the greatest of all scientific discoveries, the greatest discovery that has been made since the discovery of the controlled use of fire by prehistoric man, was the discovery of the ways in which the immense stores of energy that are locked up in the nuclei of atoms can be released.

Many scientists contributed to this discovery. Among them we may mention some of the great ones—Pierre and Marie Curie, Albert Einstein, Ernest Rutherford, Niels Bohr, Ernest Lawrence, Frederic and Irene Joliot-Curie, Otto Hahn, Enrico Fermi.

This discovery, by providing power in essentially unlimited quantities for the future world, should lead through its peaceful applications to a great increase in the standards of living of people all over the world.

It is this discovery also that has changed the nature of war in an astounding way. . . .

The Nagasaki and Hiroshima bombs had explosive energy somewhere between 15,000 and 20,000 tons of TNT. Each of them was accordingly about fifteen thousand or twenty thousand times more powerful than a one-ton blockbuster. Each was about one thousand times as powerful as the greatest of the great bombs with conventional explosives used in the Second World War. Each of them killed more than ten thousand times as many people as were killed by the average blockbuster of the Second World War. . . .

A new type of thinking is essential if mankind is to survive and move to higher levels. Today the atomic bomb has altered profoundly the nature of the world as we know it and the human race consequently finds itself in a new habitat to which it must adapt its thinking. Modern war, the bomb, and other discoveries present us with revolutional circumstances. Never before was it possible for one nation to make war on another without sending armies across borders. Now with rockets and atomic bombs no center of population on the earth's surface is secure from surprise destruction in a single attack. Should one rocket with an atomic warhead strike Minneapolis, that city would look almost exactly like Nagasaki. Rifle bullets kill men, but atomic bombs kill cities. A tank is a defense against a bullet, but there is no defense in science against a weapon which can destroy civilization.

Our defense is not in armaments, nor in science, nor in going underground. Our defense is in law and order.

Henceforth every nation's foreign policy must be judged at every point by one consideration: Does it lead us to a world of law and order or does it lead us back toward anarchy and death? I do not believe that we can prepare for war and at the same time prepare for a world community. When humanity holds in its hand the weapon with which it can commit suicide, I believe that to put more power into the gun is to increase the probability of disaster.[15]

Pauling was promoting the concepts of international peace and world law at a dangerous time, when there was renewed bitter struggle between the capitalist and the communist, or Marxist/socialist, economic systems. By the late 1940s, people who took a public stance like Pauling's and joined groups organized to promote world peace were virtually inviting themselves to be regarded and treated as subversives.

At first Pauling did not realize that by talking publicly about atom bombs and world peace he had brought on constant scrutiny of the FBI and other governmental agencies aiming not just to protect classified secrets but also to identify agitators and traitors. Later, he disregarded the risk to his reputation as an honorable U.S. citizen. But this new half-time line of work would affect both his professional and private life. He had, at least, a close companion in his dedication to peace. For more than two decades his wife had made it possible for him to focus primarily on scientific work instead of being distracted by a multitude of domestic responsibilities. Ava Helen now strongly supported—indeed, highly encouraged—his entry into social activism and politics. She accompanied him on most of his speaking and meeting ventures, particularly after their children were grown and out on their own.

Both Paulings, in any case, always followed their own ideas of what constituted public morality and a good social conscience. They were unwilling ever to back away from taking a strong stand on issues that they felt had moral meaning, especially any with deep consequence for the future of humankind. They already had a local reputation for independent thinking and action that didn't fit into current views of loyalty to the nation. For instance, in 1942 they had opposed the rounding up and mass internment of Japanese Americans during the war. Some of Pauling's students and assistants were of Japanese ancestry, and he had intervened in their behalf. The Paulings experienced public opprobrium based on prejudice and irrational patriotism. He vividly remembered it many years later when talking with Japanese philosopher Daisaku Ikeda.

BEFORE World War II, my wife hired a Japanese gardener. When the war started, all Japanese were transported to detention camps. As a consequence, we lost our gardener. But before long, someone telephoned my wife to inform her of a young Japanese American Nisei who, though already inducted into the American army, had two weeks' leave to settle family affairs and would like to take care of our garden during the interim. My wife and I belonged to a group that was protesting the treatment of Japanese people in California. Perhaps a member of that group suggested that my wife hire the young man.

She did hire him. But he worked only one day, because on the night of his having been hired, a rising sun and the words "Americans die, but Pauling loves Japs" appeared painted on our garage and mailbox.

We were threatened, and the threats grew worse after word of the incident appeared in the newspapers. I had to go to Washington, D.C., on some war work; and while I was away, the local sheriff was compelled to put a guard around our house to protect my wife.[16]

By the late 1940s the enemy's identity had changed. The Japanese no longer posed a perceived threat to national security, but communists, Marxists, and their socialist or pacifist fellow travelers did. There was an added reason for people's hesitation to speak out as Pauling did. The late 1940s and early 1950s saw the rise of the fervidly (or rabidly) anticommunist Senator Joseph McCarthy of Wisconsin and other Red-baiting politicians, such as Richard M. Nixon, who used their militant patriotism as election strategies. Mao Tse-tung and the communists' takeover of China in 1949 had confirmed Americans' fears, exacerbated by the outbreak of the Korean War. Public concern about communism mounted as a loud crescendo.

Without considering the possible consequences to his reputation, Pauling—intent upon world peace—often signed statements and petitions, lent his name to organizations, attended meetings, publicly protested loyalty oaths demanded of government employees, and promoted causes whose advocates he perhaps had not investigated at all. This was a perilous action, and for several years, from 1952 to 1954, the Passport Division of the State Department refused to reinstate Pauling's passport. Though he had repeatedly stated that he was not a communist, it was declared that his anticommunist statements were not strong enough. He was arbitrarily allowed or denied visas to travel abroad on a case-by-case basis.

In 1954, Pauling's old friend and fellow scientist J. Robert Oppenheimer, who had worked on developing the atomic bomb, was being labeled a probable communist. He had been dismissed from his position as advisor to the Atomic Energy Commission (A.E.C.). Pauling bravely came to his defense, calling this "A Disgraceful Act" in the article of that title published in *The Nation*.

THE suspension as a security risk of Dr. J. Robert Oppenheimer from his advisory activities with the Atomic Energy Commission constitutes a disgraceful act on the part of the government of the United States. This display of ingratitude toward a man who has been foremost among the scientists of the country in unselfish service to the nation cannot be justified by any rational argument. His reputation has been seriously damaged, and it will remain damaged, no matter what is the outcome of his loyalty hearing.

There is no question about Dr. Oppenheimer's loyalty. It is stated, and he has himself announced, that in the 1930s he had among his friends some who were interested in communism and social and political questions and who may have been communists. In the 1940s and 1950s he spent most of his time, energy, and extraordinary ability in outstanding service to the nation.

Throughout this recent period he has sacrificed his own career as a productive scientist in order to perform a public service. Few men have better deserved the nation's highest award to civilians, the Presidential Medal for Merit, which was presented to him by President Truman at the end of the war.

The conclusion that Dr. Oppenheimer is a loyal and patriotic American must be reached by any sensible person who considers the facts. It must have been reached by the officials of the A.E.C., and by President Eisenhower himself. We are accordingly forced to believe that the recent action is the result of political considerations—that Dr. Oppenheimer has been sacrificed by the government to protect itself against McCarthyism.

This action is sure to have serious consequences to the nation. It may be expected that many thoughtful scientists will decide that it is dangerous to make an important contribution to the national welfare, and that they should not accept employment in government agencies, or should be careful that their contributions are not outstanding.

It has been said that Dr. Oppenheimer opposed the H-bomb program at the time, 1949, when the initiation of this program was under consideration.

Dr. Oppenheimer is to be commended if he advanced moral and ethical arguments against the manufacture of that greatest of all weapons of mass destruction, the H-bomb. In 1949, when scientists knew that the bomb could be made, the government might have initiated vigorous negotiation with the rest of the world to achieve a system of general disarmament, abolition of atomic weapons, and settlement of differences between nations through arbitration and the use of a strengthened

United Nations. Now, when hydrogen bombs have been made and exploded, there no longer remains even the slightest doubt that their use in war would cause the end of civilization. Instead of raising trivial questions about Dr. Oppenheimer's loyalty, which he has demonstrated time and again since 1940 through his deeds, the government should be asking him to use his great intellectual ability, in collaboration with many other outstandingly able physical scientists, social scientists, and specialists on international relations and other aspects of the world problem, to find a practical alternative to the madness of atomic barbarism.[17]

Pauling could no longer safely discount the consequences to his reputation, whether as an objective scientist or a loyal citizen of the United States. To political conservatives, to scientists who disapproved of his agitating activities, to the administrators of organizations and institutions (including some at Caltech), and to many people within the American public who subscribed to the doctrine of "My country . . . right or wrong," he was becoming the embodiment of a foolish or even traitorous scientist.

In some respects, from the late 1940s to the mid-1950s Pauling's real work in lobbying government officials and world leaders in behalf of peace and human welfare was only just beginning. In 1952 the first hydrogen bombs were exploded by the United States; in the following year the U.S.S.R. would show that its nuclear scientists and engineers had H-bombs of their own and would test them, too. Even more than before, Pauling saw the need to examine carefully and speak out publicly against all this testing and buildup of nuclear weapons—the waste of resources and expertise, the exacerbation of Cold War conflicts, the possibly dire consequences to human health. To do so, however, would only get him in further trouble with people who supported the U.S. government's program of always staying ahead of the Soviets.

Again and again, in a somber or baleful litany, Pauling in his speeches and writings, in his interviews with journalists and with the press corps, would recite the statistics relating to megatons of explosive force. And the numbers in the potential body count inevitably grew as the weapons themselves got more powerful, especially after 1954, when America's new superbomb was exploded at Bikini atoll in the South Pacific. Pauling's article "The World Problem and the Hydrogen Bomb" asked thinking Americans to understand how urgent the need for peacemaking had become. He also began emphasizing the issue of nuclear fallout (discussed in Chapter 8).

DURING the last few months some hydrogen bombs were exploded, and information about the destructive power that they have been shown actually to possess has been released to the public.

We—everybody in the world—must consider this information, and decide what its significance is, and what must be done if the world is to be saved for posterity.

The time has come when man must show whether he is properly called "Homo sapiens," or whether he is still an unthinking brute.

First, I may mention that the hydrogen bomb itself is not the worst weapon that could be built—it is outdone as a death-dealing instrument by the cobalt bomb, which has not yet been tried out on a large scale. But the hydrogen bomb itself is enough to force us to consider the direction in which the world is moving, and to reach a decision about the future of the world. . . .

Scores of atom bombs have been detonated in the United States, South Pacific, Australia, and Russia. It has been reported that the latest models of atom bombs are around ten or twenty times as destructive as the original ones.

No announcement has been made as to the number of atom bombs that are stored up in the arsenals of the leading countries of the world, but a recent newspaper report said that the informed guesses lie in the region of five thousand atom bombs in the United States, five hundred atom bombs in Russia.

Scientists recognized early that still more destructive bombs could be made. The energy radiated by the sun arises not from the fission of the nuclei of heavy atoms, but from the fusion of light nuclei—the reaction of four protons, the nuclei of hydrogen atoms, to form a helium nucleus. . . . The ordinary atom bomb is to serve simply as the detonator of the hydrogen bomb—to raise the hydrogen to a high enough temperature to cause it to react. The reaction leads to the formation of helium, and to the evolution of energy, for about a ton of hydrogen, one thousand times as great as that for an atom bomb. . . .

I have seen a statement that the hydrogen bombs that have already been exploded in the South Pacific have an explosive power between 2 and 14 million tons of TNT; the largest one is thus described as a 14-megaton bomb, meaning equivalent to 14 million tons of TNT.

This bomb would destroy practically everything within an area of one thousand square miles—that is, within a circle about thirty-five miles in diameter. One of these bombs detonated over New York would destroy the whole city, out to the suburbs, and might kill 5 million people. One of them detonated over Los Angeles would destroy this city, perhaps killing 2 million people. . . .

Part of the danger from atom bombs and hydrogen bombs lies in the radioactive material that they produce. Japanese fishermen eighty miles from the hydrogen bomb explosion were burned by radioactive ash that fell on their ship. Many radioactive fish had to be destroyed in Japan—even fish caught by fishermen two thousand miles from the scene of the explosion. The effect of radioactivity can be very greatly increased by a simple process—just the addition of some cobalt to the atomic bomb or hydrogen bomb. . . . For 2 or 3 billion dollars enough cobalt bombs could be made to kill everybody—animals too—in the United States, simply by detonating them off the western coast, and in other suitable places, and allowing the winds to carry the products of detonation across the country. Probably some millions of people would be killed elsewhere too. . . .

There is no effective defense against these great weapons of destruction. We might hope that 50 percent of the airplanes or guided missiles carrying atom bombs or hydrogen bombs that are launched against American cities could be shot down. The most optimistic estimate that I have seen is that possibly 90 percent could be shot down. If we accept this, all that would be necessary would be for around ten or twenty missiles carrying hydrogen bombs to be launched toward New York; probably one or two of them—more probably still, several—would escape being shot down and would explode over the city. Then New York would be destroyed. While we were losing our major cities, we might be able to cause equally great damage in Russia. . . .

The atom bomb and the hydrogen bomb have become powerful weapons of destruction in the hands of powerful nations opposed to one another. If international affairs continue along the lines characteristic of the whole past history of the world, we shall sooner or later see the outbreak of a hydrogen-bomb war. No nation will benefit from such a war—it may be expected confidently that a hydrogen-bomb war, if it comes, will result in the destruction of most of the cities in the world, the death of hundreds of millions of people, the end of the present civilized world.

There is only one way in which this end can be avoided. This way is to work for peace in the world. In the past each great nation has attempted, in its diplomatic negotiations with other nations, to achieve results which benefit itself preferentially over other nations. Negotiations between nations have not in general been carried out on a high ethical plane. The representatives of a nation do not ask whether an agreement that is being made—or a declaration of war—will benefit the world as a whole, but only whether the act will benefit one's own nation. The time has now come when it is to the advantage of everybody in the world, of every nation in the world, to solve international problems in a peaceful manner—which necessarily means in a just and ethical manner —and not to solve them by force, by that ultimate resource of powerful injustice, war.

We have not yet seen either the Russian diplomats or the American diplomats attacking world problems in this way. Bluster, the issuance of ultimatums, and threats of war are still the approved methods of world diplomacy. . . .

We shall have war in the world—hydrogen-bomb war, leading to the destruction of civilization—unless the decision is made to give up war as a method of decision in international affairs.

Continued negotiation, arbitration, analysis of international problems will have to be carried out—year after year. We should be willing to spend billions of dollars every year to achieve permanent peace. . . .

We must work, through the United Nations, to find what actions we can take to meet the needs and desires of Russia, and what actions Russia can take to meet the needs and desires of the United States and other nations. There are many other ways—liberation of oppressed colonial peoples, raising the standard of living—in which effective action can be taken for the avoidance of war; we should have a Department of Peace, in Washington, to attack this problem in a broad way.

I am sure that peace can be achieved if the effort is made—and the hydrogen bomb requires that the effort be made.

During the preceding hundreds and thousands of years the world has seen the steady development of more and more powerful weapons of destruction. At each stage the opinion was expressed that war had at last become so terrible that it would have to be abandoned as a means of settling differences among nations. This belief has in the past been found to be false, even though 20 million people were killed in the First World War, 100 million in the Second World War. Now, however, we are forced to accept the conclusion that atom bombs and hydrogen bombs can destroy civilization on earth. War must be outlawed, be abandoned as a means of settling differences among the nations of the earth. The time has come for man to show that he has the power of reason, that he can behave in accordance with high ethical and moral principles, that, recognizing the universal brotherhood of man, and putting into practice the teachings of the world's great religious leaders, he can take action that will preserve civilization and humanity.[18]

Eight

The Perils of Fallout

Each nuclear bomb test spreads an added burden of radioactive elements over every part of the world. Each added amount of radiation causes damage to the health of human beings.

During the late 1940s and early 1950s, when Pauling began devoting about half his time to alerting the public to the terrible danger of nuclear weapons and the necessity of achieving cooperation and peace between the Cold War adversaries, he initially approached these subjects as a scientist. He felt as entitled and as qualified as anyone to talk about nuclear-weapons development and the establishment of international law. He also took it upon himself to question the rationality and morality of allowing a vast, costly military and industrial program to utilize and profit from a technology developed by scientists. He introduced not only the probable death statistics in nuclear attack but also, increasingly, the health hazards involved in testing such weapons.

As the age of the atom advanced, Linus Pauling said early, then often, that the creation of the atomic bomb and the growing array of nuclear weapons meant that warfare was no longer an acceptable option among nations of the world, especially those superpowers that possessed nuclear armaments. In other words, the prospect of a nuclear war should act as a deterrent. That the safety and survival of millions of people were now greatly endangered made it mandatory to be rational in resolving conflicts and to begin working continually toward universal peace. Through skillful negotiating and mutual respect, an understanding, even a partnership in sharing the world's resources, might be achieved between countries and between diverse economic systems. Diplomacy should no longer consist of the hostile blustering and bluffing that was then typical. There was too much at stake—the destruction of everyone and everything.

Pauling's interest in establishing ethical principles in society took root as the United States and then other nations developed nuclear weapons to use against one another. Intrigued with a statement that Benjamin Franklin had made in a letter to chemist Joseph Priestley in 1780, he quoted or paraphrased it innumerable times in his speeches and writings. He even chose it as the epigraph to Chapter 1 in the new edition (1957) of his textbook *General Chemistry,* and also in *College Chemistry:*

THE rapid progress true Science now makes occasions my regretting sometimes that I was born so soon. It is impossible to imagine the heights to which may be carried, in a thousand years, the power of man over matter. O that moral Science were in as fair a way of improvement, that men would cease to be wolves to one another, and that human beings would at length learn what they now improperly call *humanity.*[1]

Pauling at this time was also deliberately setting out to make a science of morality and then bring morality into government. In 1958 he articulated his views eloquently in *No More War!*

SOMETIMES I think that I am dreaming; I can hardly believe that the world is as it is. The world is beautiful, wonderful—scientists every year uncover, discover, more and more wonders of organic and inorganic nature. Man is a wonderful organism—the human body, with its millions of millions of cells, molecules of many different kinds entering into chemical reactions with one another; the human mind, capable of feats of complex calculation, of abstract reasoning infinitely beyond those of even the greatest giant electronic calculator.

Man has developed admirable principles of morality, which in large part govern the actions of individual human beings. And yet, we are murderers, mass murderers. Almost all of us, even many of our religious leaders, accept with equanimity a world policy of devoting a large part of our world income, our world resources—$100 billion a year—to the cold-blooded readying of nuclear weapons to kill hundreds of millions of people, to damage the pool of human germ plasm in such a way that after a great nuclear war our descendants might be hardly recognizable as human beings.

Does the commandment "Thou Shalt Not Kill" mean nothing to us? Are we to interpret it as meaning "Thou shalt not kill except on the grand scale," or "Thou shalt not kill except when the national leaders say to do so"?

I am an American, deeply interested in the welfare of my fellow Americans, of our great nation. But I am first of all a human being. I believe in *morality*. Even if it were possible (which it is not) to purchase security for the United States of America by killing all of the hundreds of millions of people behind the Iron Curtain without doing any harm to anyone else, I would not be willing that it be done.

I believe that there is a greater power in the world than the evil power of military force, of nuclear bombs—there is the power of *good*, of *morality*, of *humanitarianism*.

I believe in the power of the human spirit. I should like to see our great nation, the United States of America, take the lead in the fight for good, for peace, against the evil of war. I should like to see in our cabinet a Secretary for Peace, with a budget of billions of dollars per year, perhaps as much as 10 percent of the amount now expended for military purposes. I should like to see set up a great international research program involving thousands of scientists, economists, geographers, and other experts working steadily year after year in the search for possible solutions to world problems, ways to prevent war and to preserve peace.

During the past hundred years there have been astounding developments in science and technology, developments that have completely changed the nature of the world in which we live. So far as I can see, the nature of diplomacy, of the conduct of international affairs, has changed very little.

The time has now come for this aspect of the world to change, because we now recognize that the power to destroy the world is a power that cannot be used.

May our great nation, the United States of America, be the leader in bringing *morality* into its proper place of prime importance in the conduct of world affairs![2]

Pauling felt that scientists such as himself could determine the codes for a good life in society as well as anyone—perhaps even better. He did not believe that contemporary elected or appointed officials had the exclusive right to make decisions that affected people's well-being. Pauling was influenced by older protesters such as Einstein and Bertrand Russell. But soon he provided the energetic leadership for a campaign to halt nuclear-weapons testing, especially aboveground, which was already contaminating the earth's atmosphere, soil, and water—the grave consequences of which might not be seen for years.

As part of his public-education campaign, Pauling often patiently explained to listeners and readers basic information about radiation, as he did in these selections from his popular book *No More War!* (1958).

ALL high-energy radiation—alpha rays, beta rays, gamma rays, X rays —produces ions when it passes through matter. The number of ions produced depends on the quantity of radiation and upon its nature. The ions are produced by the interaction of the radiation with atoms or molecules; electrons are knocked out of the atoms or molecules, leaving the atoms or molecules with a positive electric charge—these electrically charged atoms or molecules are called ions.

The roentgen is defined as the amount of radiation that produces a certain number of ions when it passes through a cubic centimeter of air.

The biological effects of high-energy radiation seem to be produced in large part by the power of the radiation to produce ions, and accordingly these biological effects are pretty well represented quantitatively by the roentgen as a unit of measurement of the radiation.

Although the alpha particle has great energy, it does not penetrate very far through matter. The alpha particles from radium on the surface of the skin or in the bones of a human being penetrate through a distance of only about one-fiftieth of an inch into the adjacent or surrounding tissues.

Other radioactive materials emit beta rays. Beta rays are electrons that are shot out of the atomic nucleus at high speed. Radioactive nuclei such as strontium 90 and yttrium 90 emit beta rays that have the power of penetrating about one-eighth of an inch through the tissues of the human body.

Also, radioactive nuclei often emit gamma rays. The gamma rays from radioactive nuclei are in general highly penetrating—these rays can pass completely through the human body, irradiating all parts of it, in the same way as X rays from an X-ray tube operated at very high voltage.[3]

The decomposition of a radioactive atomic nucleus has an interesting characteristic—it is a chance event, which cannot be precisely predicted. A radon nucleus freshly formed by decomposition of a radium nucleus does not start out in life in full vigor, gradually get old, and then, after it reaches old age, die with the emission of an alpha particle and conversion into a polonium nucleus. Instead, it has exactly the same chance of decomposing during the first second after its birth from the radium nucleus as it has after the tenth second or after the hundredth second or the thousandth second.

It is found that after a sample of radon has been watched for 3 days, 19 hours, and 12 minutes, half of the radon nuclei have decomposed. After another 3 days, 19 hours, and 12 minutes, half of the remainder have decomposed, and so on. The activity falls off slowly, with half of the remaining nuclei decomposed after each period of time of this length.

This period of time, the time required for half of the nuclei to undergo radioactive decomposition, is called the half-life of the radioactive material.

Radioactive elements differ very much in their half-lives. The half-life of radium is 1,600 years—much greater than that of radon. The half-life of the parent element uranium 238 is 4,500 million years. Uranium is accordingly an element whose nuclei are almost but not quite stable against radioactive decomposition. Many other nuclei are completely stable, so far as can be detected. Some radioactive nuclei have very short half-lives, only a small fraction of a second.[4]

Pauling especially focused on the potential hazard from carbon 14.

ORDINARY carbon is carbon 12—its nuclei are made of six protons and six neutrons. The nuclei of radiocarbon, carbon 14, are made of six protons and eight neutrons. Carbon 14 is produced at a steady rate in the upper atmosphere by a reaction of nitrogen nuclei caused by cosmic rays. The atoms of carbon 14 in the course of time combine with oxygen to form carbon dioxide, and the carbon dioxide, radioactive and nonradioactive alike, is absorbed by plants, which build the carbon atoms into their tissues. Animals that eat the plants also build the carbon atoms into their tissues.

It is unfortunate that the bomb tests have caused the concentration of carbon 14 in the atmosphere to increase by about 10 percent, and it now continues to increase at the rate of 2 or 3 percent per year.[5]

Pauling presented some of the consequences of radiation when people were exposed to it. During warfare they suffer radiation sickness. Long-term exposure to lower amounts of radiation also pose severe health threats.

THE amount of full-body radiation that causes death is somewhere between 300 roentgens and 600 roentgens. The amount 450 roentgens is estimated to cause death by radiation sickness in 50 percent of the recipients.

Many of the fatalities in Hiroshima and Nagasaki were the result of radiation sickness.

Radiation sickness, though important in the discussion of nuclear warfare, has little importance for the bomb tests. Except for the few accidental deaths that have occurred in connection with the development of nuclear weapons and the possible case of one Japanese fisherman on the fishing boat *The Fortunate Dragon*, no deaths from acute radiation sickness have been reported in the nuclear weapons programs.

The effect of a large amount of radiation on the human body in causing death from acute radiation sickness is somewhat similar to the effect of a lethal dose of an ordinary poison—perhaps closer to that of a mixture of ordinary poisons.

But radiation has other effects that are different from the effect of an ordinary poison. These effects may result from even very small exposures to radiation.

A person may swallow an amount of a poison one-thousandth of the lethal dose without doing himself any harm whatever. He may repeat this small dose day after day for one thousand days without harming himself; whereas if he had taken the one thousand small doses at one time he would have been killed. For example, he may take one sleeping pill each night for one thousand nights in succession without harm to himself, although he would have died if he had swallowed the thousand pills at one time.

But the properties of radiation are different. It is known that exposure of the reproductive organs of animals to 100 roentgens of radiation causes the same amount of damage to the genes, the same number of mutations, when the exposure is made in many small doses spread over a long period of time as when it is made in a single dose. It is probable also that many small doses of radiation are as effective in causing leukemia and bone cancer and some other diseases as a single large dose of an equal amount of radiation.

A very small amount of an ordinary chemical poison may do no harm whatever to a person; but a very small amount of radiation may harm him in such a way as to cause him to die or to have a seriously defective child.

We can understand this strange and terrible effect of radiation by considering the way in which radiation interacts with the molecules that compose the human body.

The rays of high-energy radiation are like little bullets that shoot through the body. They tear electrons away from molecules, and through subsequent reactions of the molecular ions that are formed the molecules may be broken in two, some atoms may be torn away from them, some new molecules may be formed.

The exposure of the human body to 1 roentgen of radiation causes about a thousand ions to be formed in each cell of the human body—one thousand new molecules, perhaps poisonous, in each cell. The dose of 500 roentgens that usually leads to death by acute radiation sickness causes about five hundred thousand changed molecules to be formed in each cell.

Some of the new molecules that are produced by the radiation are poisonous. If enough of them are made in the body of a human being he will die in a few days. He has died of acute radiation sickness.

In most of the cells of the human body there are, among the billions of molecules of many different kinds, a few very important ones. These are the molecules, probably molecules of deoxyribose nucleic acid, that govern the behavior of the cell, that control the manufacture of other molecules and the process by which the cell divides to form new cells, and that determine the nature of the children who are born to the person.

If one of these special molecules happens to be damaged by a single little bullet of radiation from a single radioactive atom, it may be changed in such a way as to cause the cell to divide much more rapidly than the other cells of the body. This cell may then produce a colony of rapidly dividing cells, which in the course of time would outnumber the normal cells of that type. Then the human being may die from cancer—perhaps leukemia, bone cancer, some other kind of cancer—caused by the single radioactive atom that produced the single little bullet of radiation.

This is the reason why the radioactive atoms, such as strontium 90 and cesium 137, that are being spread all over the world by the bomb tests are harmful. *There is no safe amount of radiation or of radioactive material. Even small amounts do harm.*

There is no doubt that the small amounts of radiation received by every human being from cosmic rays, natural radioactivity, and fallout radioactivity cause gene mutations that increase the number of defective children born in future generations. Geneticists agree on this, and their estimates of the number of defective children caused by a given small amount of radiation are reasonably consistent. . . .

Ionizing radiation in amounts of the order of hundreds of roentgens or thousands of roentgens causes cancer and other diseases. Within a year after the discovery of X rays it was noticed that radiologists and technicians working with X-ray tubes often developed burns on the hands, forearms, and face, which in some cases developed into cancer of the skin.[6] . . .

Pauling then detailed the various types of cancer, such as leukemia, aplastic anemia, and bone, lung, and thyroid cancers, that frequently afflicted researchers and industrial workers exposed over time to X rays and other forms of high-energy radiation, even at relatively low levels, before precautions were taken to limit their exposure or to shield them from damage. Another risk was initiated when physicians routinely used X rays or other types of radiation to treat or diagnose relatively mild health conditions, unaware of the potential in cumulative roentgens. His conclusion:

THERE is no doubt whatever that ionizing radiation produces cancer and other diseases when the exposure is large. The argument that we have given above, based on the effect of the tiny bullets in changing the important molecules that control the cells, suggests that it also has a proportionate effect in producing some cases of these diseases even when the amount is small.[7]

In the mid-1950s Pauling began focusing on what had become for him the crucial issue in these bombs—the radioactive fallout coming from the tests conducted with them. The tests took place in relatively remote regions; residents were simply removed from the vicinity, from their own lands. But the blasts hurled dust and other particles skyward that were then carried by wind and clouds, eventually to descend to earth from the stratosphere. Fallout—contaminating air, ground, oceans, and other bodies of water—posed new health risks to people and their future progeny. Whether during nuclear warfare or bomb testing, people's exposure to radioactivity, at locations far distant from explosion, could still have health consequences similar to those from incautious occupational or medical radiation—or worse. Pauling was alarmed not only by the prospect of a nuclear war, but also by the probable long-term effects of sustained testing.

WE are the custodians of the human race. We have the duty of protecting the pool of human germ plasm against willful damage.

If the human race survives the present era of nuclear peril, damage will continue to be done to the pool of human germ plasm by the radioactive elements created in the nuclear test explosions that have been carried out during the past few years and present year.

The most menacing radioactive substance of all of those produced in the bomb tests is carbon 14. It is a great menace because it lives so long—over eight thousand years—and because it is produced in large amounts—about 16 pounds per megaton of explosive energy.

Also, there is no such thing as a clean bomb with respect to carbon 14. It is made by both fusion and fission. Some of the neutrons from the bomb—fission alone, or fission-fusion, or fission-fusion-fission—escape and combine with nitrogen nuclei of the air to form carbon 14.

The entire atmosphere, the ocean, and the biosphere (plants and animals) used to contain about 160,000 pounds of carbon 14, produced in the upper atmosphere by cosmic rays. During the past four years the amount in the atmosphere has been increased by 10 percent because of the bomb tests. If the bomb tests are continued at the 1958 rate, the amount in the atmosphere will be doubled in a few years.[8]

Pauling began condemning nuclear fallout separately from the larger issue of arms development and stockpiling as first the United States, then Britain and the Soviet Union, developed superbombs and began testing them. It was only a question of time before France and the People's Republic of China, and other nations as well, would follow suit. But Pauling's campaign tactics of educating people about the health consequences of fallout made an impact. He and other antinuclear militants apparently embarrassed the governments, concerned about public opinion, into halting their tests.

In the fall of 1959, for example, Pauling gave a speech at New York's Carnegie Hall, at a meeting of the National Committee for a Sane Nuclear Policy. (It was later printed in *Mainstream* as "Fallout: Today's Seven-Year Plague.") Among the many points Pauling made in his talk given to like-minded people was his discussion of the Geneva Conferences on Bomb Tests and his observation that no nuclear tests had been conducted since November of the previous year. Certainly Pauling's efforts to stir up world opinion against testing had contributed to this lull in fallout production.

For a while Pauling felt hopeful of progress since public questioning of the health consequences, morality, and practical need for continuing nuclear tests mounted sufficiently to push the three prominent nuclear nations—the United States, Britain, and the Soviet Union—into announcing that each would, even without the enactment of a formal treaty, stop atmospheric tests. Although mutual talks had taken place, agreement could not be made regarding an inspection system—declared to be an essential part of a formal accord. Neither the peace that Pauling desired nor the self-imposed nuclear test bans were to last.

The temporary atmospheric-testing moratorium that began in 1959 ended after France began testing its own new nuclear weapons in early 1961. The major nuclear powers felt free to resume tests—which in August the U.S.S.R. announced it would do, shortly afterward setting off a new series of atmospheric nuclear tests in September. Several days later the United States resumed underground tests. The informal test ban had ended. Pauling sent telegrams of protest to Premier Nikita Khrushchev and to President John F. Kennedy. Both national leaders ignored his plea—though the Russian premier sent Pauling a long explanatory letter.

Pauling had a dark vision indeed of the world after a nuclear apocalypse, which at times seemed imminent. He gave a lurid warning in the article "The Dead Will Inherit the Earth," published in *Frontier* in November of 1961. He started by dismissing the protective abilities of fallout shelters. Americans were being encouraged by the government and the media to build small underground hideouts. Made of concrete and stocked with several weeks' worth of water, food, and other provisions, they were supposed to protect their habitants from nuclear blast and radioactive debris. In a number of speeches and articles at the time, Pauling let it be known that surviving nuclear war was dangerous wishful thinking, as he described the devastation that would take place at such a time.

WHAT are we to believe about fallout shelters? Will the construction of fallout shelters on a large scale save the lives of a great number of American people?

My answer to this question is that a great program of shelter construction would not save the lives of Americans, but would instead increase our danger.

Nobody knows exactly what would happen in a nuclear war, but many people, such as scientists of the Rand Corporation, have worked hard in an effort to make predictions about the nature of nuclear wars of various kinds. The statements that I make in this article are largely based upon the reported results of these studies.

My statements, it will be seen by the reader, differ greatly from many of the statements that have been published.

My own estimate is that a great nuclear war in which a major part of the nuclear weapons now stockpiled by the United States and the U.S.S.R. was used would result in the death of 170 million of the 180 million Americans within sixty days, and that the surviving 10 million would also soon die. If a great program of shelter construction were to be initiated, I estimate that the nuclear war that could be expected after completion of the shelter program would result in the death within sixty days of the same fraction of the American people, and the probable death of the 10 million sixty-day survivors within a year. The United States would cease to exist as a nation, and the American people would all be dead. The U.S.S.R. would cease to exist as a nation, and the Russian people would all be dead. Probably most of the other European nations would cease to exist, and the European people would be dead.

The construction of fallout shelters would not protect the American people or preserve our nation.

I believe that the goal of general and complete world disarmament can be achieved, and that it is the duty of every human being to use his time and energy and money to assist in the fight to achieve this goal, rather than in a vain effort to alleviate the consequences of nuclear war, as by the construction of fallout shelters.[9]

Finally, in August of 1963, after five years of intermittent talks among diplomats, military experts, and nuclear scientists, the superpowers agreed to forgo on-site inspections, since any sizable aboveground, underwater, and outer-space nuclear explosions would be apparent. Only underground testing was expected to continue, discreetly. After all, tests producing nuclear fallout in the atmosphere had been the greatest concern of the world's citizenry, thanks in good part to Pauling's vigorous publicizing of their health hazards.

On October 10, 1963—the date when the first international limited test-ban treaty went into effect—Pauling experienced a justifiable triumph: his efforts to halt nuclear testing in the atmosphere had actually earned him the world's highest possible recognition for humanitarian service. It was awarded, belatedly, for 1962. The Nobel Peace Committee finally found a deserving nominee.

Linus Pauling's lecture when receiving the Nobel Peace Prize in December 1963 in Oslo repeated with variations many of the talks he had already given as well as statements made to reporters for almost two decades. Still, "Science and Peace" provides a stirring and succinct record of the thinking of Pauling and others who were concerned about nuclear issues from 1945 to the time in 1963 when a limited test-ban treaty to end aboveground, or atmospheric, nuclear testing was signed. Here is the first portion, which laid the foundation for Pauling's discussion of the need for the test-ban treaty.

I believe that there will never again be a great world war—a war in which the terrible weapons involving nuclear fission and nuclear fusion would be used. And I believe that it is the discoveries of scientists upon which the development of these terrible weapons was based that are now forcing us to move into a new period in the history of the world, a period of peace and reason, when world problems are not solved by war or by force, but are solved in accordance with world law, in a way that does justice to all nations and that benefits all people.

I remind you that Alfred Nobel wanted to invent "a substance or a machine with such terrible power of mass destruction that war would thereby be made impossible forever." Two-thirds of a century later scientists discovered the explosive substances that Nobel wanted to invent —the fissionable substances uranium and plutonium, with explosive energy 10 million times that of Nobel's favorite explosive, nitroglycerine, and the fusionable substance lithium deuteride, with explosive energy 50 million times that of nitroglycerine.

The first of the terrible machines incorporating these substances, the uranium 235 and plutonium 239 fission bombs, were exploded in 1945, at Alamogordo, Hiroshima, and Nagasaki. Then in 1954, nine years later, the first of the fission-fusion-fission superbombs was exploded, the 20-megaton Bikini bomb, with energy of explosion one thousand times greater than that of a 1945 fission bomb. This one bomb, the 1954 superbomb, contained less than one ton of nuclear explosive. The energy released in its explosion was greater than that of all of the explosives used in all of the wars that have taken place during the entire history of the world, including World War I and World War II.

Thousands of these superbombs have now been fabricated; and today, eighteen years after the construction of the first atomic bomb, the nuclear powers have stockpiles of these weapons so great that if they were to be used in war hundreds of millions of people would be killed, and civilization itself might not survive the catastrophe.

Thus the machines envisaged by Nobel have come into existence, and war has been made impossible forever.

The world has now begun the metamorphosis from its primitive period of history, when disputes between nations were settled by war, to its period of maturity, in which war will be abolished and world law will take its place. The first great stage of this metamorphosis took place only a few months ago—the formulation by the governments of the United States, Great Britain, and the Soviet Union, after years of discussion and negotiation, of a treaty banning the testing of nuclear weapons on the surface of the earth, in the oceans, and in space, and the ratification and signing of this treaty by nearly all of the nations in the world.

I believe that the historians of the future may well describe the making of this treaty as the most important action ever taken by the governments of nations, in that it is the first of a series of treaties that will lead to the new world, from which war has been abolished forever.

**The Nobel speech then recapitulated the basis for the world scientists'
Bomb-Test Appeal, which Pauling had masterminded.**

ON May 15, 1957, with the help of some of the scientists in Washington University, St. Louis, I wrote the Scientists' Bomb-Test Appeal, which within two weeks was signed by over 2,000 American scientists and within a few months by 11,021 scientists, of forty-nine countries. On January 15, 1958, as I presented the appeal to Dag Hammarskjöld as a petition to the United Nations, I said to him that in my opinion it represented the feelings of the great majority of the scientists of the world.

The Bomb-Test Appeal consists of five paragraphs. The first two are the following:

We, the scientists whose names are signed below, urge that an international agreement to stop the testing of nuclear bombs be made now.

Each nuclear bomb test spreads an added burden of radioactive elements over every part of the world. Each added amount of radiation causes damage to the health of human beings all over the world and causes damage to the pool of human germ plasm such as to lead to an increase in the number of seriously defective children that will be born in future generations.

Let me say a few words to amplify the last statement, about which there has been controversy. Each year, of the nearly 100 million children born in the world, about 4 million have gross physical or mental defects, such as to cause great sufferings to themselves and their parents and to constitute a major burden on society. Geneticists estimate that about 5 percent, or two hundred thousand a year, of these children are grossly defective because of gene mutations caused by natural high-energy radiation—cosmic rays and natural radioactivity, from which our reproductive organs cannot be protected. This numerical estimate is rather uncertain, but geneticists agree that it is of the right order of magnitude.

Moreover, geneticists agree that any additional exposure of the human reproductive cells to high-energy radiation produces an increase in the number of mutations and an increase in the number of defective children born in future years, and that this increase is approximately proportional to the amount of the exposure.

The explosion of nuclear weapons in the atmosphere liberates radioactive fission products . . . [which] are now damaging the pool of human germ plasm and increasing the number of defective children born.

Some people have pointed out that the number of grossly defective children born as a result of the bomb tests is small compared with the total number of defective children, and have suggested that the genetic damage done by the bomb tests should be ignored. I, however, have contended, as have Dr. Albert Schweitzer and many others, that every single human being is important, and that we should be concerned about every additional child whom our actions cause to be born to live a life of suffering and misery. In a broadcast on July 26, 1963, President Kennedy said: "The loss of even one human life, or the malformation of even one baby—who may be born long after we are gone—should be of concern to us all. Our children and grandchildren are not merely statistics towards which we can be indifferent." . . .

It is my estimate that about 100,000 viable children will be born with gross physical or mental defects caused by the cesium 137 and other fission products from the bomb tests carried out from 1952 to 1963, and 1,500,000 more, if the human race survives, with gross defects caused by the carbon 14 from these bomb tests. In addition, about ten times as many embryonic, neonatal, and childhood deaths are expected—about 1,000,000 caused by the fission products and 15,000,000 by carbon 14. An even larger number of children may have minor defects caused by the bomb tests; these minor defects, which are passed on from generation to generation rather than being rapidly weeded out by genetic death, may be responsible for more suffering in the aggregate than the major defects. . . .

Moreover, it is known that high-energy radiation can cause leukemia, bone cancer, and some other diseases. . . .

We may now ask: At what sacrifice is the atmospheric test of a single standard 20-megaton bomb carried out? Our answer, none the less horrifying because uncertain, is: the sacrifice, if the human race survives, of about five hundred thousand children, of whom about fifty thousand are viable but have gross physical or mental defects; and perhaps also of about seventy thousand people now living, who may die prematurely of leukemia or some other disease caused by the test.

We may be thankful that most of the nations of the world have, by subscribing to the 1963 treaty, agreed not to engage in nuclear testing in the atmosphere. But what a tragedy it is that this treaty was not made two years earlier! Of the total of 600 megatons of tests so far, three-quarters of the testing, 450 megatons, was done in 1961 and 1962. The failure to formulate a treaty in 1959 or 1960 or 1961 was attributed by the governments of the United States, Great Britain, and the Soviet Union to the existing differences of opinion about methods of inspection of underground tests. These differences were not resolved in 1963; but the treaty stopping atmospheric tests was made. . . .

The remaining paragraphs of the Bomb-Test Appeal of 1957 read as follows:

So long as these weapons are in the hands of only three powers an agreement for their control is feasible. If testing continues, and the possession of these weapons spreads to additional governments, the danger of outbreak of a cataclysmic nuclear war through the reckless action of some irresponsible national leader will be greatly increased.

An international agreement to stop the testing of nuclear bombs now could serve as a first step toward a more general disarmament and the ultimate effective abolition of nuclear weapons, averting the possibility of a nuclear war that would be a catastrophe to all humanity.

We have in common with our fellow-men a deep concern for the welfare of all human beings. As scientists we have knowledge of the dangers involved and therefore a special responsibility to make those dangers known. We deem it imperative that immediate action be taken to effect an international agreement to stop the testing of all nuclear weapons.

How cogent is this argument? Would a great war, fought with use of the nuclear weapons that now exist, be a catastrophe to all humanity?

Consideration of the nature of nuclear weapons and the magnitude of the nuclear stockpiles gives us the answer: it is Yes.

Pauling's lecture also summed up the current statistics as to the dire consequences of nuclear warfare.

THE world's stockpile of military explosives has on the average doubled every year since 1945. My estimate for 1963, which reflects the continued manufacture of nuclear weapons during the past three years, is 320,000 megatons. . . .

The significance of the estimated total of 320,000 megatons of nuclear bombs may be brought out by the following statement: If a 6-megaton war were to take place tomorrow, equivalent to World War II in the power of the explosives used, and another such war the following day, and so on, day after day, for 146 years, the present stockpile would then be exhausted—but, in fact, this stockpile might be used in a single day, the day of World War III. . . .

A major nuclear war might well see a total of 30,000 megatons, one-tenth of the estimated stockpiles, delivered and exploded over the populated regions of the United States, the Soviet Union, and the other major European countries. . . .

[It is estimated] that sixty days after the day on which the war was waged 720 million of the 800 million people in these countries would be dead, 60 million would be alive but severely injured, and there would be 20 million other survivors. The fate of the living is suggested by the following statement by Everett and Pugh:

> Finally, it must be pointed out that the total casualties at sixty days may not be indicative of the ultimate casualties. Such delayed effects as the disorganization of society, disruption of communications, extinction of livestock, genetic damage, and the slow development of radiation poisoning from the ingestion of radioactive materials may significantly increase the ultimate toll.

Pauling's lecture, drawing to its conclusion, picked up a dominant theme in his activism: how the invention of nuclear warfare made the achievement of world peace and the reign of world law essential.

Is there no action that we can take immediately to decrease the present great danger of outbreak of nuclear war through some technological or psychological accident or as the result of a series of events such that even the wisest national leaders could not avert the catastrophe?

I believe that there is such an action, and I hope that it will be given consideration by the national governments. My proposal is that there be instituted with the maximum expedition compatible with caution a system of joint national-international control of the stockpiles of nuclear weapons. . . .

Even a small step in the direction of this proposal, such as the acceptance of United Nations observers in the control stations of the nuclear powers, might decrease significantly the probability of nuclear war.

There is another action that could be taken immediately to decrease the present great hazard to civilization. This action would be to stop, through a firm treaty incorporating a reliable system of inspection, the present great programs of development of biological and chemical methods of waging war, . . . [which would] have enormous lethal and incapacitating effects on man and could also effect tremendous harm by the destruction of plants and animals. Moreover, there is a vigorous effort being made to develop these weapons to the point where they would become a threat to the human race equal to or greater than that of nuclear weapons. The money expended for research and development of biological and chemical warfare by the United States alone has now reached $100 million a year, an increase of sixteenfold in a decade, and similar efforts are probably being exerted in the Soviet Union and other countries.

To illustrate the threat I may mention the plans to use nerve gases that, when they do not kill, produce temporary or permanent insanity, and the plans to use toxins, such as the botulism toxin, viruses, such as the virus of yellow fever, or bacterial spores, such as of anthrax, to kill tens or hundreds of millions of people.

The hazard is especially great in that, once the knowledge is obtained through a large-scale development program such as is now being carried out, it might well spread over the world, and might permit some small group of evil men, perhaps in one of the smaller countries, to launch a devastating attack.

This terrible prospect could be eliminated now by a general agreement to stop research and development of these weapons, to prohibit their use, and to renounce all official secrecy and security controls over microbiological, toxicological, pharmacological, and chemical-biological research. Hundreds of millions of dollars per year are now being spent in the effort to make these malignant cells of knowledge. Now is the time to stop. When once the cancer has developed, and its metastases have spread over the world, it will be too late.

The replacement of war by law must include not only great wars but also small ones. The abolition of insurrectionary and guerrilla warfare, which often is characterized by extreme savagery and a great amount of human suffering, would be a boon to humanity.

There are, however, countries in which the people are subjected to continuing economic exploitation and to oppression by a dictatorial government, which retains its power through force of arms. The only hope for many of these people has been that of revolution, of overthrowing the dictatorial government and replacing it with a reform government, a democratic government that would work for the welfare of the people.

I believe that the time has come for the world as a whole to abolish this evil, through the formulation and acceptance of some appropriate articles of world law. With only limited knowledge of law, I shall not attempt to formulate a proposal that would achieve this end without permitting the possibility of the domination of the small nations by the large nations. I suggest, however, that the end might be achieved by world legislation under which there would be, perhaps once a decade, a referendum, supervised by the United Nations, on the will of the people with respect to their national government, held, separately from the national elections, in every country in the world. . . .

In working to abolish war we are working also for human freedom, for the rights of individual human beings. War and nationalism, together with economic exploitation, have been the great enemies of the individual human being. . . .

We, you and I, are privileged to be alive during this extraordinary age, this unique epoch in the history of the world, the epoch of demarcation between the past millennia of war and suffering and the future, the great future of peace, justice, morality, and human well-being. We are privileged to have the opportunity of contributing to the achievement of the goal of the abolition of war and its replacement by world law. I am confident that we shall succeed in this great task; that the world community will thereby be freed not only from the suffering caused by war but also from hunger, disease, illiteracy, and fear through the better use of the earth's resources, the discoveries of scientists, and the efforts of mankind; and that we shall in the course of time be enabled to build a world characterized by economic, political, and social justice for all human beings, and a culture worthy of man's intelligence.[10]

What Pauling said in his Nobel Peace Prize lecture was far more general than personal, just as his Nobel Prize in Chemistry lecture had been almost a decade earlier. Pauling told little indeed about his own important part in many of the antinuclear and peace-promoting activities mentioned. Nor did he mention the terrible social, political, and even economic fallout that had descended upon him during this challenging yet highly stressful period in international relations, when his own life had been both stressed and challenged. Among the many unpleasant experiences during the course of his antinuclear campaign that Pauling did not mention in his Nobel speech was his subpoena by a Senate Internal Affairs Subcommittee, chaired by Republican senator Thomas Dodd of Connecticut. Pauling was interrogated about his method of collecting the signatures of scientists around the world on the petition presented in January of 1958 to the United Nations. The subpoena came in 1960, more than two years after the event. The central purpose supposedly was to permit the subcommittee to identify the organizations that had assisted Pauling in this effort—for it was believed that many were influenced or run by communists. The course of the proceedings indicated that, more specifically, the committee wanted to know the names of key individuals who assisted Pauling by circulating the petition among their acquaintances—an act they obviously interpreted as intended to undermine the U.S. government's military power. When Pauling declined to name his petition helpers, the subcommittee gave him time to think it over, with the implication that if he did not comply he would be in contempt of Congress and subject to imprisonment.

Although by 1960 the demagogue Joseph McCarthy had left the Senate in public disgrace, his accusatory spirit remained. The notorious Hollywood blacklist reigned, terrorizing the entertainment industry, and the FBI under J. Edgar Hoover's direction gathered adverse reports about anyone engaged in civil rights activities or suspected of leftist-leaning connections. Pauling was a prime target.

When Pauling first faced the subcommittee, he had let them know why he refused to turn over the asked-for names.

THE circulating of petitions is an important part of our democratic process. If it were to be abolished or greatly inhibited, our nation would have made a step toward deterioration—perhaps toward a state of dictatorship, a police state.

I am very much interested in our nation, the United States of America, and in the procedures that were set up in the Constitution and the Bill of Rights. Now, no matter what assurances this subcommittee might give me about the use of the names of the people who circulated the petition that I wrote, I am convinced that these names would be used for reprisals against these believers in the democratic process, these enthusiastic, idealistic, high-minded workers for peace. I am convinced of this because I myself have experienced the period of McCarthyism and to some extent have suffered from it, in ways that I shall not mention. I am convinced of it because I have observed the workings of the Committee on Un-American Activities of the House of Representatives and of this Subcommittee on Internal Security of the Judiciary Committee of the Senate. I feel that if these names were to be given to this subcommittee the hope for peace in the world would be dealt a severe blow. Our nation is in great danger now, greater danger than ever before. The danger is not from peace or from the workers for peace, or from the circulators of petitions urging international law and international agreements. It is from the stockpiles of nuclear weapons that exist in the world, which have the capability of destroying the world. . . . This danger, the danger of destruction in a nuclear war, would become even greater than it is now if the work for peace in the world, peace and international law and international agreements, were hampered.

A terrible attack is being made now in the United States on the efforts of our government to achieve international agreements on stopping the bomb tests and for disarmament. This attack is being made by representatives of defense industries who benefit financially from the Cold War. . . .

If two nations, the U.S.S.R. and the United States, were to try to follow the policy of each striving to exceed the other in the power of destruction year after year, then we would truly be doomed to catastrophe. Then the nuclear war that will destroy civilization surely would come.

I believe that the work for peace and morality and justice in the world needs to be intensified now, and I plan to do whatever I can in working for peace in the world, working for international agreements about disarmament; always, of course, such that they increase the safety of the United States and do not decrease it, and involving controls and inspection. . . .

I am responsible for my actions, and I wrote the petition and I sent it out to people, asking that they get signatures to it. I have selected the many people to whom these petitions were to be sent. I think that my reputation and example may well have led these younger people to work for peace in this way. My conscience will not allow me to protect myself by sacrificing these idealistic and hopeful people, and I am not going to do it. As a matter of conscience, as a matter of principle, as a matter of morality, I have decided that I shall not conform to the request of this subcommittee.[11]

Pauling's courage and words were admired by many Americans, even some who before then had criticized him for his antinuclear activities. He managed to avoid incarceration for contempt of Congress, in good part because of the furor stirred up by the subcommittee's inquiry, which resembled a witch-hunt. It brought an avalanche of protests to Washington—irate letters, telegrams, and phone calls—and numerous editorials in the press. This experience was only one of hundreds of public and private affronts to Linus Pauling during the years when he incessantly crusaded against nuclear weaponry: its creation and testing, its deployment on land, sea, and in the air, the frequent threats to use it when provoked—indeed, its very existence in the world. From the start he had allies in this long campaign, but for some years his detractors held the power and largely manipulated the press into portraying him as either a traitor or a goofy, maybe mad, scientist. (For instance, *Life* magazine greeted his receipt of the Nobel Peace Prize as a "weird insult from Norway.") He had the compensatory satisfaction, at least, that he and Ava Helen had thousands of supporters around the nation and elsewhere in the world; the number grew as the years went by and people began comprehending the insanity of it all.

When the Nobel Peace Prize Committee had announced its choice of Linus Pauling on October 10, 1963, the day the partial test-ban treaty took effect, Pauling got the news of the award via the forest rangers' telephone line while staying at his remote coastal ranch.

IN 1963, I was notified that I had received the Nobel Peace Prize for 1962. Several reporters then asked, "Of these two Nobel Prizes, which one do you value the more?" I said, "Well, I was happy to receive the Nobel Prize for Chemistry, but I had just been having a good time carrying out studies in the field of chemistry and trying to make discoveries. The Nobel Peace Prize is the one I value more because it means there is a feeling that I have been doing my duty to my fellow human beings. And perhaps it means that working for peace has become respectable. They won't take away my passport anymore or call me up before a congressional committee and threaten me with jail anymore. So it is the Nobel Peace Prize about which I feel happier." [12]

Telegrams of congratulations poured in to Pauling from around the world. Yet at home the prophet was shown little honor. Upon returning to Pasadena, he received only tepid congratulations from most Caltech administrators and from faculty members other than his friends and advocates—when he got them at all. No large celebratory event was planned to honor him, as had happened in 1954 with the Nobel Prize in Chemistry and as was customary for faculty members in far less honorific circumstances. The chemistry division ignored him, so finally the biologists put a party together for him.

To be sure, Caltech-raised or -based researchers had won various Nobel Prizes in the past, unshared or shared, in chemistry, physics, and physiology or medicine. But this one for peace seemed strange and out of place—particularly because Pauling's involvement with social activism had been a great embarrassment to the administration, who had to respond to protests from conservative trustees and alumni who contributed considerable financial support to the institute. In fact, President Lee DuBridge had forced Pauling to resign as chemistry division chairman in 1957 because of his activities. Now, when told of Pauling's award, DuBridge remarked to the *Los Angeles Times* that "it's very remarkable that a person should get a second Nobel Prize, but there's much difference of opinion about the value of the work that Professor Pauling has been doing."

Two weeks later, Pauling announced that he would depart from Caltech after finishing up any research and teaching obligations for the remainder of the academic year. It seemed an appropriate time for him to move to an environment that would actually welcome his involvement in world peace issues while also accommodating his liking for theoretical research. A new chapter in Linus Pauling's life and career would soon begin, and he would be encouraged to pursue his deep commitments to peace and humanitarianism as a scientist in society.

Nine

Apostle of Peace

This great organism, humankind, is now master of the earth, but not yet master of itself.

The 1962 Nobel Peace Prize validated Linus Pauling's position as a world statesman for peace. In the three decades of his life afterward he felt properly credentialed to speak out during various antiwar and other public-activist campaigns that enlisted his support. Having two unshared Nobels made him unique—as if doubly worth heeding. Pauling's presence and words were now more in demand, better respected. He was even invited now to some places where he had not been welcome before.

Shortly after returning to the United States with his second Nobel gold medal, Linus and Ava Helen announced their plan to leave Pasadena. Pauling said publicly that he was taking a leave of absence from his position as professor and research director at Caltech to join the staff of researchers at the Center for the Study of Democratic Institutions (CSDI) in Santa Barbara. Noted educator-scholar Robert M. Hutchins, retired from the University of Chicago, headed the organization, a subsidiary activity of the Fund for the Republic. Its stated mission was "to promote the principles of individual liberty expressed in the Declaration of Independence and the Constitution. Its studies are chiefly directed at discovering whether and how a free and just society may be maintained under the strikingly new political, social, economic, and technological conditions of the second half of the twentieth century." Pauling was sure that his words and actions as antinuclear agitator and peace promoter would be far better appreciated at CSDI than at Caltech. The Paulings bought a small house in Santa Barbara and settled in by the summer of 1964.

In *No More War!* Pauling had proposed a world peace research organization. Now at CSDI he returned to this idea and proposed ways for it to be funded: out of the superpower nations' huge military budgets. It would be "a cheap insurance policy," he said.

Pauling's involvement with Hutchins' scholarly think tank was appropriate for someone with a firm, idealistic belief in democracy. A free thinker, Pauling believed that others could and should be too—and would be if they received enough encouragement, information, and guidance. Though increasingly he recommended the application of the scientific method of thinking to solve societal problems of many kinds, he also was never reluctant to use words weighted with philosophical and ethical implications, such as justice and morality.

Perhaps nowhere did Pauling make a stronger statement of his own moral principles and innate humanitarianism than in his talk at the international conference sponsored by CSDI in New York in February of 1965 to discuss the late Pope John XXIII's eloquent encyclical *Pacem in Terris.* An anthology of the assembled scholars' contributions, *To Live as Men: An Anatomy of Peace,* contained Pauling's statement. He had said some of the same things earlier. But in places here he took a somewhat more theological and spiritual tone, in keeping with the liberal-inclined pope's religious language.

I accept, as one of the basic ethical principles, the principle of the minimization of the amount of suffering in the world.

I do not accept the contention that we cannot measure the suffering of other human beings, that we do not know what is good and what is evil.

Even though my relationship to myself is subjective and that to other human beings is objective, I accept the evidence of my senses that I am a man, like other men; I am "fed with the same food, hurt with the same weapons, subject to the same diseases, healed by the same means, warmed and cooled by the same winter and summer"; when I am pricked I bleed, as do other men; when I am tickled, I laugh; when I am poisoned, I die. I cannot contend that it is the result of anything but chance that I am I, that this consciousness of mine is present in this body; I cannot in good faith argue that I deserve a better fate than other men; and I am forced by this logic to accept as the fundamental ethical principle the Golden Rule: "As ye would that men should do to you, do ye also to them likewise" (Luke).

I know what causes me to suffer. I hope that other human beings will take such actions as to keep my suffering to a minimum. And it is my duty to my fellow men to take such actions as to keep their sufferings to a minimum.

We suffer from accidents, from natural catastrophes, from disease, from the ills accompanying the deterioration of age, and also, in a sense the most viciously, from man's inhumanity to man, as expressed in economic exploitation, the maldistribution of the world's wealth, and especially the evil institution of war. . . .

During year after year, decade after decade, century after century, the world has been changed by the discoveries made by scientists and by their precursors—by those brilliant, original, imaginative men and women of prehistoric times and of more recent times who learned how to control fire, to cook food, to grow crops, to domesticate animals, and then to build wheeled vehicles, steam engines, electric generators and motors, and nuclear fission power plants. And, of course, in the early days, the scientists were the theologians, the religious leaders, too. Sometimes the thought occurs to me that the world will not be saved unless we return to this condition. . . .

The scientists of the East and West are, so far as I can see, very much like one another. They resemble one another not only in their knowledge of science but also in their acceptance of moral principles. It seems to me when I compare scientists with diplomats—with other people—that the scientists of the whole world are more closely related to one another than scientists are to other people in their own country. There is a better understanding among them than with other people. This understanding must spread. The discoveries that scientists have made provide now the possibility of abolishing starvation and malnutrition, and improving the well-being, and enriching the lives of all of the world's people.

The effect of the discoveries of scientists in decreasing the amount of human suffering is illustrated by the control that has been achieved over the infectious diseases. In many parts of the world it is now rare for women to die of puerperal infection, for infants to die of diphtheria or scarlet fever, for people to die of diseases such as smallpox or bubonic plague. Cancer remains a cause of great human suffering not yet brought under control; but we may hope that this terrible disease will also succumb in the next few decades to the attack on it that is being made by scientists.

The results of medical discoveries and technological developments have not yet been made available to all of the world's people. Modern means of waging war seem to be more easily available to the underdeveloped countries than drugs, food, and machines for increasing the production of goods.

Our system of morality as expressed in the operating legal, social, and economic structures is full of imperfections, and these imperfections have been accentuated during recent decades. There is great misery caused by the abject poverty of about half of the world's people; yet most of the scientists and technologists of the world today are working to make the rich richer and the poor poorer, or are working on the development and fabrication of terrible engines of mass destruction and death whose use might end our civilization and exterminate the human race. . . .

I believe that it is a violation of natural law for half of the people of the world to live in misery, in abject poverty, without hope for the future, while the affluent nations spend on militarism a sum of money equal to the entire income of this miserable half of the world's people.

Pope John in his great encyclical letter said that every human being is a person, that every man has the right to life, to bodily integrity, to food, clothing, shelter, rest, medical care, and social services, to security in case of sickness, inability to work, widowhood, old age, unemployment, or deprival otherwise of the means of subsistence through no fault of his own; the right to respect for his person, to his good reputation, the right to freedom in searching for truth and in expressing and communicating his opinions; the right to be informed truthfully about public events; the right to share in the benefits of culture; the right to a basic education and to suitable technical and professional training; the right to free initiative in the economic field and the right to work under good working conditions, with a proper, just, and sufficient wage; the right to private property, with its accompanying social duties; the rights of residence and of freedom of movement, of membership in the human family and membership in the world community.

Most human beings are now denied these rights. It is our duty to work to achieve them for everyone. . . .

One of the most evil aspects of human suffering is the absence of any justice or meaning in its distribution. Accidents, natural disasters, and excruciatingly painful diseases strike with the blind malevolence of chance. . . .

The injustice and immorality of the great wars of the past would be far transcended by a great war in the nuclear age, a war in which the devastating weapons involving nuclear fission and fusion that now exist were used. Instead of tens of millions, hundreds or even thousands of millions of human beings might be killed. Great nations might be exterminated. Civilization might come to an end. There is even the possibility that the human race would not survive the catastrophe.

As rational and moral beings, we are forced now to find a rational and moral alternative to war.

As he was increasingly doing in his speeches and writings, Pauling decried his own nation's dominant role in the Vietnam War.

A T this critical time in history I cannot omit discussion of the terrible and dangerous crisis in Southeast Asia.

The people in this part of the world have suffered from war and oppression for thousands of years. For more than two centuries great European powers exercised colonial control over the people. Then, nineteen years ago, there began the bloody revolution of the Indo-Chinese war.

After eight years the effort to make use of reason and morality in place of military might and the sacrifice of human lives came close to success. Britain, France, and the Netherlands terminated their rule over their Asiatic colonies. And now—and these facts are omitted in most recent official statements—I must say that one great country, my own country, the United States of America, although present at the Geneva Conference of 1954, refused to sign the final agreement to bring an end to war in Vietnam, and then, in 1956, together with the new government that had been imposed upon South Vietnam, refused to allow the people of South Vietnam to select their own government by ballot, as required by the Geneva Accords.

This repudiation of the principles of arbitration and negotiation and of the principles of democracy has led to ten years of savage guerrilla warfare and to retaliation with helicopters, fire bombs, chemical defoliation and destruction of crops, the uprooting of hundreds of thousands of peasants from their homes and their forced concentration into strategic hamlets, great air raids with modern fighter bombers, and the threat of nuclear destruction through attack from Polaris submarines lying close offshore. And now there impends the horrible danger of escalation to the catastrophe of a civilization-destroying nuclear war.

Now the time has come to obey the exhortation of Pope John: the exhortation to cease military aggression, to bring this evil war to an end, to meet and negotiate and make a great practical application of the principles of morality and justice.[1]

Although Pauling was an agnostic or atheist, after quoting Pope John XXIII's benediction at the end of his encyclical message imploring peace on earth and goodwill to men, he had enough reverence for life, human and otherwise, to offer his own responsive blessing: "I join in this prayer, and I express now my hope and my belief that we shall succeed in banishing from the earth forever the great immorality of war."

Though involved with visions of peace, Pauling never forgot science. In a delightful part of a discourse to an assemblage of science teachers, entitled "The Social Responsibilities of Scientists and Science," Pauling commented on the illegality of teaching evolution in public schools within the states of Tennessee and Arkansas and then described in some detail his recent molecular-evolution work with his young French colleague Emile Zuckerkandl, in which they had counted the number of similarities and differences in the amino-acid residues in cytochrome C among various animal species and speculated on their implications.

THERE is as yet no similar information about human beings and gorillas with respect to cytochrome C. But the amino-acid sequence in the cytochrome C of human beings is known, and also that in the cytochrome C of the rhesus monkey. Now, of the amino-acid residues in these 104 positions, the same choice among 20 has been made by man and the rhesus monkey in the first position, in the second position, the third, the fourth, the fifth, and so on down. Only one position out of the 104 involves a different amino acid for the rhesus monkey and human beings.

It may surprise them in Tennessee and Arkansas when this information gets there. I can predict what the result will be when someone gets some gorilla cytochrome C.

With the horse, there are eight differences. This is because the evolutionary strains that led to human and horse separated from one another about eight times as far back in history as did the strains that led to present-day strains for human and rhesus monkey. There are 21 differences between tuna fish and humans, so the separation between man and tuna fish occurred even earlier in the scale of evolution.

With moth cytochrome C there are 30 differences and 74 identities. With bread yeast and humans, there are about 45 amino acids that are different and about 59 that are identical. Think how close together man and this other organism, bread yeast, are. What is the probability that in 59 positions the same choice out of 20 possibilities would have been made by accident? It is impossibly small. There is, there must be, a developmental explanation of this. The developmental explanation is that bread yeast and man have a common ancestor, perhaps 2 billion years ago. And so we see that not only are all men brothers, but men and yeast cells, too, are at least close cousins, to say nothing about men and gorillas or rhesus monkeys. It is the duty of scientists to dispel ignorance of such relationships.[2]

Then Pauling became more serious.

SCIENTISTS also have the duty to help educate those of their fellow citizens who represent what C. P. Snow called "the other culture." If I remember correctly, Lord Snow divided the two cultures in this way: the scientists, the people who understand the world, including our scientific knowledge of it, constitute one culture, and the nonscientists, who understand only those parts of the world that we describe as nonscientific, constitute the other culture. Not long ago Professor Denis Gabor of the Imperial College in London participated in a symposium in the Center for the Study of Democratic Institutions. After this experience with a large group of social scientists, he suggested that the division is not really between the scientist and the nonscientist, but between those people interested in facts and ideas on the one hand, and those interested in words on the other.

This seems harsh, but I think that there is something to it. I recall a series of discussions at the center about the presidency, in which presidents of the United States were categorized as Washingtonian, or Jeffersonian, or Hamiltonian. Finally I asked, "Have you laid down some attributes of the actions or decisions of the presidents and assigned to them their percentage weights of Washingtonianism, Jeffersonianism, and Hamiltonianism, and then analyzed the actions and decisions of the various presidents to find out what the quantitative conclusions are? Have you done this, or have you done anything such as to lead you to think that two different people who classified the presidents among your three categories would reach the same conclusions about them?" The answer was that this sounded interesting but had not been done. Yet, this is what they must do—not just use words that they don't define and carry on vague discussions, but try to make their concepts more precise, to have ideas that can be closely related to fact.

In the class of people who are interested in facts and ideas, we have, of course, most scientists, and also a good number of nonscientists who think along the same lines even though they don't have scientific training. In the other class—those interested in words—we have some scientists and some philosophers, and many nonscientists. I remember reading a book on philosophy in which the author went on, page after page, on the question: If there is a leaf on a tree and you see that it is green in the springtime and red in fall, is that the same leaf or is it a different leaf? Is the essence of leafness still in it? Words, words, words, but "chlorophyll" and "xanthophyll"—which are sensible in this connection of what has happened to that leaf—just don't appear at all. Admittedly, we have some people who are called scientists who are in the category of those who talk about words rather than facts and ideas.

What is the solution going to be? I believe that the ultimate solution will be that *everyone* will have a knowledge of science, but it will take a generation, two generations, for us to reach this goal, even in the United States. I believe that we *shall* reach this goal if the world is not destroyed. I believe that reason will win out and that the world will continue to improve.[3]

At the end of his lecture, Pauling enunciated a powerful social credo for scientists that he would maintain for the rest of his lifetime—his last words echoing those in his Nobel Peace Prize lecture.

W E, as scientists, have the general social responsibilities resulting from our knowledge and understanding of science and its relation to the problems of society. It is not our duty to make the decisions, to run the world. It is, rather, our duty to help educate our fellow citizens, to give the benefit of our special knowledge and understanding, and then to join with them in the exercise of the democratic process.

Among the problems with which we may be concerned are the pollution of the atmosphere, the pollution of water supplies, fluoridation of water and use of other public health measures, contamination of the earth with pesticides, with lead from leaded gasolines, misuse of chemicals as food additives, the location of nuclear power plants in thickly populated centers, the best use of scientific and medical knowledge to decrease the amount of human suffering caused by poverty and disease, and especially the prevention of the destruction of civilization by nuclear war.

I believe that we shall succeed in abolishing war, in replacing it by a system of world law to settle disputes between nations, that we shall in the course of time construct a world characterized by economic, political, and social justice for all human beings and a culture worthy of man's intelligence.[4]

In the following year, at the Noranda Lectures at Expo 67 in Canada, Pauling talked at length on the topic "Science and the World of the Future." Here was his conclusion:

W E are achieving a mastery of the physical world and the biological world. The practical world of business and politics lags behind; the policies in this practical world remain the antiquated ones of the last century, determined by individual, corporate, and national selfishness. I believe that the time has come for the nations of the world to cooperate, for national patriotism to be replaced by loyalty to the whole of mankind, for war to be abandoned and replaced by world law based upon the principles of justice and morality, for the resources of the world to be used for the benefit of the whole of mankind; the time has come for us to take steps and to make such decisions, individually, in groups, in business, in government, as will have the greatest estimated probability of decreasing the amount of human suffering in the world.[5]

Pauling made various excursions abroad on dual missions: to deliver major speeches on peace-connected topics and to attend scientific conferences. He did not arrive empty-handed at his destination but would come armed with cogent facts about such topics as regional history, agriculture, population, religious conflicts, and the natural environment, so that he might offer suggestions for solving particular national problems. For instance, in 1967 Pauling gave two major memorial lectures, the Chettyar in Madras and the Azad at New Delhi. After briefly discussing his philosophy of life (appropriately giving a DNA twist to the Hindu notion of reincarnation), he expressed grave concern about the causes and consequences of overpopulation. Manipulating statistics like a skilled demographer, he gave India as a notable example of what lay in store for the future world in a sadly diminished quality of life. (His Azad lecture was published afterward in India as a book, *Science and World Peace*.) In later years Pauling sometimes declared that worldwide population expansion had overtaken his nuclear fears as the most urgent problem to be solved by humanity. Not only would it have a devastating effect on the environment, but excessive population would greatly increase the probability of mass suffering—with or without war.

Linus Pauling never returned to regularly teach and conduct research at Caltech—despite originally announcing an extended leave of absence. (Eventually, with the changing of the old guard, Pauling was treated with homage during special occasions of his return to campus, such as celebrating his eighty-fifth and ninetieth birthdays. And in 1990 the Linus Pauling Chair in Physical Chemistry was created.)

In 1967 he took a second leave of absence, this time from CSDI. Pauling missed the synergistic camaraderie of fellow scientists, the ambience of intense exchange in scientific symposia, and the precision of laboratory research in the physical and biological sciences.

He accepted an offer to join the faculty at the University of California, San Diego. His duties were light and the administrators' expectations of him were small, permitting him a great deal of traveling and talking elsewhere. He remained there for only two academic years—assuming a

proffered full-tenured professorship at Stanford University in the autumn of 1969. Pauling took his billowing anti–Vietnam War wrath with
him up to Palo Alto, where he frequently took part in student demonstrations.

The 1960s and early 1970s were proving to be a difficult time—rest-
less, turbulent, transformational—for the United States. The civil rights
movement, the Vietnam War protests, the sexual revolution, the dropout
and drug culture, and other matters involving free-speech radicalism,
mayhem, and occasional violence: all tore at the social and intellectual
fabric of the nation. Pauling—who in many respects was a classic Puritan
—disapproved of some of these activities. Yet he felt in tune with the
rebellious spirit within this latest generation of youth, particularly those
who were attending college. Their antiwar stance especially stirred his
fighting spirit for peace. This is the Pauling who cheered them on, all
the while preaching nonviolence—in this speech protesting the invasion
of Cambodia in 1970.

NIXON ordered the invasion. Nixon has gone too far!

There is no doubt now—if Nixon is not stopped, we are doomed—
doomed to continually burgeoning militarism, to a military takeover in
this country, to ever-increasing world strife, and finally, ultimately, to
the world-destroying cataclysm of nuclear war.

There is madness in the White House, paranoia—the madness that
comes from the recognition of the possession of unlimited power—the
power to destroy the whole world, to destroy civilization, to wipe out the
human race—power that never existed before. This madness must be
replaced by sanity.

We deserve to have a good government, a moral government, a government of which we can be proud, a government that will foster and
cherish our wonderful young people and not send them off to be slaugh-
tered or send them to prison, a government that will strive to provide
proper food and housing and medical care and education for every
person, that will not force young men into bondage, into the immorality
of killing their fellow men.

Nixon could control the economy by an excess-profits tax. But he
wants the poor to pay for the war and the Pentagon—so he raises interest
rates and increases unemployment.

I say: America belongs to the people of America, not just to the exploiters, the unconscionable rich, the militarists. This great country
needs an administration that will work for the well-being of all the people
—not just for the militarists and the war profiteers.

I am proud of our young people, who have taken the lead. Our young men who go to jail for their moral principles are the heroes of our country. They show us the path of duty and morality.

Now, the time has come when we must revolt against our oppressors, our exploiters, against the immoral crew in our national government. Stop the war in Vietnam! Stop the attack on Cambodia! Pull the CIA out of Vietnam, Cambodia, Laos, the whole of Southeast Asia!

Pull out the ground troops, the advisors, the jet bombers, the battle-ships! Abandon the dictators, the puppet government! Turn Vietnam over to the Vietnamese people![6]

It was with some pride that in 1970 Pauling accepted the International Lenin Peace Prize. In his speech at the Soviet embassy in Washington, D.C., he shared in the mutual relief that "we have survived this decade!"

THE outlook for world peace looks dark in 1971. There are wars in Southeast Asia and the Middle East—unreasonable, irrational, immoral wars, unworthy of man, who has progressed so far in his investigations of the physical and biological world, who prides himself on his intellect, and who considers himself to be moral and humane.

Hundreds of billions of dollars are being wasted on war and militarism. One-tenth of the resources and labor of the United States and corresponding amounts in other countries are being thrown away—worse than being thrown away: are being used for the killing, maiming, and burning of men, women, and children, for laying waste the countryside, destroying crops, villages, cities, schools, hospitals. And how much work is being devoted to efforts to end the wars by political means to decrease the burden of militarism? A miniscule amount, a few million dollars, a few score of men negotiating!

But I have hope. In 1960 scientists who had studied the problem estimated the probability of a cataclysmic nuclear war during the 1960–1970 decade as 50 percent. And we have survived this decade!

I believe that the probability of nuclear war is now smaller, that we may well have passed the nadir.

I can understand that the politicians have found it hard to move from the prenuclear period of reliance on war into the nuclear age, the age of international cooperation and continuing world peace. It has been said that the quantum theory was fully accepted only when the older generation of physicists had died off. The twenty-five years since 1945 has not been quite enough for the old politicians to die off, but they are on their way out; and the young people have shown us that we have good reason to have hope for a better world of the future.

I hope that the war in Southeast Asia will soon be brought to an end.

I hope that the Strategic Arms Limitation Talks will soon move forward, providing a treaty that will permit the reduction of the military budgets of both the U.S. and U.S.S.R. by several tens of millions of dollars per year.

I hope that the U.S. and the U.S.S.R. will then continue to cooperate in actions to favor world peace, to decrease the amount of human suffering. One such action, in which other developed nations should join, would be to forbid the provision, by gift or sale, of jet bombers and fighters, guns, tanks, ammunition, and other means of making war, to the less developed countries of the world, and instead, in collaboration with the United Nations, work to achieve a just resolution by negotiation of the disputes that arise.

I am confident that this cooperation of the U.S. and the U.S.S.R., and ultimately also the Chinese People's Republic, will come about, that it will lead to a series of epoch-making treaties, to the development of an effective system of world law, based upon the principles of justice and morality, and ultimately to a world of permanent peace, in which the resources of the earth and the products of man's labor are used for improving the well-being of the whole of humankind.[7]

In 1974 Pauling made his fourth move since leaving Caltech in 1963— to his own institute. Pauling's participation in political activism became more intermittent. He never again found a peace-connected cause to engage him almost wholly. One explanation for this is that Pauling moved his crusading spirit elsewhere—into nutritional medicine (Part IV). He would encounter innumerable entrenched adversaries there— individuals, associations, and government agencies—just as he had in his previous campaigns. Many of the old tactics served him again on this new front line.

By 1975 the Vietnam War had shuddered to its ignoble close. In that year the Paulings were invited to attend the Albert Schweitzer Centennial Birth Anniversary Commemoration Symposium on "Reverence for Life and the Way to World Peace," held in Japan. Linus Pauling prepared a brief statement, "We Must Throw Off the Yoke of Militarism to Achieve Albert Schweitzer's Goal for the World."

THE goal of all humanists is to achieve a world in which the resources and the products of man's labor are used in such a way as to permit every person to lead a good life. This goal involves consideration not only of the human beings now living but also of those of future generations, for thousands or tens of thousands of years. As Albert Schweitzer emphasized, we must have reverence for life; not just for mankind, but also for other forms of life. We must also have reverence for the wonderful inanimate aspects of the world. It is our duty to preserve the world's resources and the world's wonders for the benefit and enjoyment of future generations.

We have recognized only rather recently that the resources of earth are finite, and that we are in danger of exhausting them. Also, there is clearly a limit to the number of people that can be enabled to live a good life. The fact that the lives now lived by most of the people on earth are unsatisfying and impoverished indicates that we have already exceeded the optimum population for the earth.

The well-being of mankind is determined by the available natural resources and by man's labor. If the resources are in considerable part wasted, as in war and militarism, the people suffer. If large numbers of people are out of work, decreasing the productivity of man's labor, the people suffer.

At the present time there is a depression, with many people out of work, and inflation, diminishing the purchasing power of the money that is available to them. It is irrational to have a shortage of goods, leading to high prices, and at the same time to have many people out of work. The social and economic system needs to be changed in such a way as to avoid such calamities. Inflation is especially hard on older people, who have retired and are living on fixed income. Inflation is a way of robbing them of their share of the world's wealth. During inflation, as during other times, the rich get richer and the poor get poorer.

It is war that is responsible for most of our ills—war and militarism. About 10 percent of the world's wealth, amounting to $250 billion per year, is spent on military expenditures. The war in Vietnam, which resulted in no significant achievement except that of bringing about the deaths of hundreds of thousands of people and injury to many more, cost more than $250 billion. Even though that war has come to an end, the military budget in the United States has continued to increase. Great efforts continue to be made toward improving nuclear weapons and their delivery systems and developing new weapons of mass annihilation.

We must work to get the great nations to agree on a limitation of armaments, a decrease in military budgets, the prevention of the spread of nuclear weapons to more and more nations, and other measures that will in the course of time lead to permanent world peace and to the goal of a world worthy of the dignity of man, in which every person has the opportunity to lead a good life, to the full extent of his capabilities.[8]

Ava Helen Pauling added her own shorter but matching statement:

Peace in the world requires that we recognize the real enemies of mankind and work to vanquish them. These enemies are war, ignorance, disease, poverty, hunger, the pollution of the environment, the wasting and ruination of the world's resources by militarism. These enemies of mankind and of world peace are the imperatives on which we must focus our energies and our abilities if we wish to secure the survival of man and the survival of the planet Earth.[9]

Like other venerable scientists (Einstein readily comes to mind), Linus Pauling was often regarded as a philosopher who could advise others about how best to think and live. Like an oracle, he was also consulted at propitious times—the start of a decade or an even more major milestone—and asked to look into the future. For instance, when the nation's bicentennial year arrived in 1976, Pauling composed two important statements. The first one he delivered at hearings conducted by the Senate Committee on Government Operations on the issue "Our Third Century: Directions." On February 6 Linus Pauling asked the committee members, "What should be our goals?" He then responded with a prepared text, first explaining his moral principle of the minimization of suffering, and then going into such issues as chronic unemployment, overpopulation, the uneven distribution of wealth, and, of course, the unwise allocation of funds to war and militarism.

One of the subhead topics in his discourse to the senators was "Our Responsibility to Future Generations."

I believe that it is our moral duty to take such actions as will leave the world in a condition such that future generations of human beings are able to lead good and full lives, with the minimum of suffering. We cannot, of course, be sure about the future, but we should strive to predict the probable course of mankind and the world, and then make our decisions in such a way as seems most likely to minimize the suffering not only of people now living and their immediate descendants but also of future generations, even those in the distant future, with, of course, smaller weight given to the future because of the uncertainty in our predictions.

One of the sources of happiness for many people is the enjoyment of the forests, wildernesses, deserts, lakes, and other natural wonders, and the many different kinds of animals and plants. We should not deprive the people of future generations of this source of happiness, and we should accordingly conserve these wonders, and not permit them to be sacrificed to the advance of agriculture and technology.

Future generations should also not be deprived of their share of the world's natural resources. Some of these resources are inexhaustible. An example is sunlight. We may convert some of the energy of sunlight into electric power, which, after doing useful work, is finally converted to heat and warms the earth, as it would have done if it had not been converted to electric power. The sun will continue to shine for future generations, whether we use it for electric power or not. Solar energy, like energy obtained from the winds or tides, does not rob future generations.

Coal and oil, however, are present on earth in finite amounts, and when they are burned the supply is decreased. These are valuable substances, not only as fuels but also as the raw materials for making drugs, textiles, and many other compounds of carbon. But we are depleting their stocks at such a rate that they may be exhausted in a few generations.

Use of nuclear fission for generation of power is not the solution, because there is only a limited supply of fissionable elements, and our extensive development of fission power plants would soon exhaust the nuclear fuel, leaving little or none for future generations. A thorough study of the energy problem, with consideration of our obligations to future generations, would probably lead to the conclusions that great efforts should be made to obtain energy from the inexhaustible sources, and also that the amount of energy used should be kept as small as possible. There is no doubt that in past decades an increase in the amount of energy available has led to an improvement in the quality of life for many people, but there also is no doubt that a large fraction of the energy used is now wasted. Many common devices, such as air conditioners, are now built to operate inefficiently, using much more power than necessary. . . .

What number of people living in the United States would permit those individual human beings to lead the best lives, the fullest lives, the richest lives, with the greatest freedom from the suffering that is caused by poverty, malnutrition, and disease, and with the greatest satisfaction of their intellectual curiosity, of their cultural needs, and of their love of nature?

The people in the United States do not have the happiness that comes from drinking good natural water; instead they drink dilute sewage containing chlorine and organic and industrial contaminants. They breathe air contaminated with oxides of nitrogen and sulfur, with hydrocarbons and aldehydes, with lead and carbon monoxide and soot. There is increasing encroachment on the national parks, the great forests, and the wilderness regions, and increasing damage to the wildlife. The quality of the food continues to deteriorate. I believe that we should have in the United States such a number of people that every person could lead a full and rich life and such as would lead to the preservation of the beauties and wonders of the world for the enjoyment of future generations.

My analysis led me to the estimate 150 million for this optimum population of the United States. I urge that the Congress arrange that a continuing study of this question be made, that the value of the optimum population be agreed on, and that steps be taken to achieve this goal. [10]

Pauling could not, however, provide specific proposals to these legislators for how the federal government could shrink the American populace. Malthus's three ghastly means of population control—war, famine, and disease—were wholly unacceptable to him.

Then in April of 1976 Linus Pauling gave the American Chemical Society's own centennial address, on the topic "What Can We Expect for Chemistry in the Next One Hundred Years?" It repeated some of his message to the Senate committee but added material more appropriate to his chemist listeners. Both speeches took place during the nation's vaunted two-hundredth-birthday celebration. Perhaps to counterbalance some of the chauvinistic patriotism of America's bicentennial year, however, Pauling allowed a pessimistic tone into his message, particularly at the conclusion—despite his preferred tendency to see the future in a hopeful, even cheerful way.

So far I have talked about what our goals should be. I am an optimist, and I think that by the year 2076 war will have been almost entirely eliminated from the world, and replaced by a system of world law. The expenditures on militarism will be much less than at the present time. The social and economic systems will have changed . . . in such a way that in the United States, the Soviet Union, and other countries, systems will have been developed that incorporate the good points of the present system and eliminate some of their bad points; for example, great unemployment that now causes so much suffering will have been eliminated. More use will be made of man's labor, and less use of energy, in doing the world's work. . . .

The population of the world will have become stable a century from now. I am not able to predict its value. We have just reached 4 billion people, and it is likely that there will be 8 billion people in the world by the year 2010. My estimate of the optimum population for the world is 1 billion. By 2076 we may well be approaching that limit, from above. . . .

During the coming century chemists and other investigators will, I believe, succeed in finding the regimens, nutritional and environmental, that would lead to a decreased rate of aging and increased life expectancy. If civilization survives, there will be, I predict, an increase by twenty-five years (to about age one hundred) in the life expectancy, with a corresponding increase in the length of the period of well-being, before the deterioration that culminates in death.

But what is actually going to happen during the next century? We see that the leaders of the nations and the government in general, as well as the people, concentrate on the immediate problems. It is unusual for a country to develop a five-year plan, and unheard of to have a hundred-year plan. Only when a crisis arises, when a catastrophe occurs, do governments take action. I am afraid that within twenty-five or fifty years there will occur the greatest catastrophe in the history of the world. It might well result from a world war, which could destroy civilization and might well be the end of the human race, but civilization might end because of the collapse of the systems upon which it depends. Paul Ehrlich has pointed out that the collapse could take many forms, the complete loss of oceanic fisheries through overfishing, marine pollution, and the destruction of estuaries, which could lead to global famine. Or the end of civilization might result from weather changes induced by governments to improve the yield of crops, or it might end by the rapid destruction of the ozone layer, or by the accumulation of poisonous wastes that would make the air unbreathable and water unpotable.

I am forced, as I think about what has happened in the world during my lifetime and as I observe governments in their processes of making decisions, to conclude that the coming century is probably going to be one in which the amount of suffering reaches its maximum. Unless we are wiser than we have shown ourselves to be in the past, we chemists and we citizens, we advisers of the government and we government officials, there will be a catastrophe during the coming century, perhaps a series of catastrophes. The human race might survive. By 2076, we shall, I hope, have solved these problems, and from then on we may have a world in which every person who is born will have the opportunity to lead a good life.[11]

Pauling's intense involvement with peace issues gradually declined—particularly after the death of Ava Helen in 1981, since she had inspired and accompanied him in his work for international peace. The influence that she wielded in developing her husband's social conscience and then in encouraging and supporting his political activism is incalculable. In most of these ventures she had been at his side—or even out on her own.

At a conference in British Columbia in 1983, "The Prevention of Nuclear War," Pauling, by then age eighty-two, gave a long and discursive speech entitled "The Path to World Peace"—a favored title at that time. It shows some of the fascination and charm in Pauling's tendency to reminisce and ramble off to slightly irrelevant matters, especially as he got older. Yet it made him seem accessible and personable.

I love this world. I have had a good life. In 1923 I got married to a fine young woman whom I had seen in the chemistry class [I taught] at Oregon Agricultural College. She was the smartest girl in the class, also by far the best looking! Perhaps even more important, for some reason when she saw me she took a liking to me, and of course I wasn't able to resist!

I have been fortunate in many ways, and have been able to see many of the wonders of the world. I like everything about the world. I like the mesons and the hadrons, and the electrons and the protons and the neutrons; and the atoms, the molecules, the self-replicating molecules; the microorganisms, the plants and animals; the minerals; the zunyite and cuprite, and pyrite and marcasite and andalusite, and all of the other minerals; the oceans and the mountains, and the forests; the stars and the nebulae and the black holes out there; the Big Bang 18 billion years ago. I like all of it!

I like satisfying my intellectual curiosity. As a scientist, I have been fortunate to have been aware of what has been done during the last [century]. Every month I read about something new and interesting about the universe that some scientist has discovered, such as what it was that caused the extinction of the dinosaurs 64 million years ago. It is really wonderful, the world, and one wonderful part about it is that there are sentient beings here who are able to appreciate the wonders of the world, to understand them!

So I feel that we have a duty to try and prevent the nuclear war, to reverse this situation. I believe that it can be done. Otherwise I wouldn't be here. Why should I waste my time if the effort is not going to be successful? I might as well be enjoying myself, making some quantum mechanical calculations! . . .

The only solution is to have peace in the world, continued perpetual peace. It will take a long time to achieve this goal, but we have to be starting on it now. You must not think that individuals are unimportant. Each individual human being can contribute something. Of course, enough individuals can have a great effect. . . . People as a whole, especially the young people, can get the government to change its policy. You know how to do it, to demonstrate, to vote whenever there is a possibility of voting. I do not advocate taking violent action of any sort. I am not in favor of harming people or destroying property. As for civil disobedience, I do not advise people one way or the other. If someone feels strongly enough to be willing to go to jail because of his ethical principles, then he might want to commit civil disobedience. After all, Bertrand Russell went to jail as a young man protesting against the First World War and he went to jail as an old man protesting against nuclear weapons.

Then Pauling described to his listeners a vivid scenario of what might happen to humanity if a nuclear holocaust occurred.

In a nuclear war in which, let us say, 30,000 megatons were used, a [potential toll might] amount to 300 *billion deaths*. That is 75 times the population of the earth! The 60,000 megatons is 150 times the population of the earth. So sometimes I say that we have a seventy-five-fold overkill capability, or really we have a 150-fold overkill capability.

Of course, conditions would not be the same everywhere. Some people live out in the country and might escape death, but the results of studies are much the same as one performed back in 1959 by Hugh Everett III and George E. Pugh of the Institute of Defense Analysis in Washington. In testimony before a Senate hearing, they described the effect of a 10,000-megaton attack on the United States and a similar attack on the Soviet Union.

Sixty days after the attack, of 800 million people in the United States and Europe (including the European part of the Soviet Union), 720 million would be dead, 60 million severely injured, and 20 million more not yet dying but waiting to die. They would have to deal with the disorganization of society, disruption of communications, extinction of livestock, genetic damage, the slow development of radiation poisoning from the ingestion of radioactive materials, and many other problems. In the rest of the world, fallout would cause a tremendous amount of damage to hundreds of millions of people.

Other problems of a great nuclear war only recently have been subjected to careful analysis, as additional scientific information has been obtained about chemical reactions involved in the production and destruction of ozone molecules and the stratospheric ozone layer. A 5,000-megaton war might destroy 70 percent of the ozone layer. By extrapolation, a 20,000-MT war might destroy 90 percent. With even a 70 percent destruction of the ozone layer, so much ultraviolet light would get through that human beings might not be able to survive. The ozone layer would take many years to heal.

A nuclear war would start fires over much of the Northern Hemisphere, and the Southern Hemisphere also, to the extent that nuclear bombs were exploded there. Almost all forests, houses, and other combustible materials would burn. The amount of smoke and dust thrown up from groundbursts might be enough to prevent sunlight from getting to the surface of the earth for months. We know that species of animals and plants die out under this condition.

There was an earlier great extinction for which human beings were not responsible. It occurred 64 million years ago, whereas human beings separated from the anthropoid apes only 4 million years ago. Sixty-four million years ago a very interesting phenomenon occurred. The paleontologists studying layers of rock find that there is a series of layers containing a great many fossils deposited over millions of years. Above these are more layers deposited over millions of years but nearly free of fossils. There is a much smaller number of fossils, representing a lesser variety of animals. These rocks are observed all over the world. The explanation of this fact is that 64 million years ago an asteroid 11 kilometers in diameter hit the surface of the earth. It is believed that it did not hit the continent mass but landed in the ocean, because it would have left a visible crater had it hit a continent. When the asteroid hit the rocks at the bottom of the ocean, it was vaporized by the energy of the impact, and it vaporized also ten times its mass of rock. Once up in the atmosphere, this vapor condensed to dust, and it is thought that for a year or two sunlight did not get to the surface of the earth. Plants stopped growing, plankton in the ocean stopped growing, many species of plants died, and many animals also died. More than half of the existing species of plants and animals died out in this great extinction, including all eighteen species of dinosaur, which had been the predominant life form for 150 million years.

We know that this happened because of a very interesting geological structure discovered by Professor Louis Alvarez and his son, Walter. All over the earth one finds these strata separated by a layer of clay 2 centimeters thick. This layer of clay is found even in cores obtained from the ocean bottom. This is the dust created by the impact of the asteroid, which settled down in a few years to the surface of the earth. There is one hundred times as much iridium in this clay as there is in terrestrial rocks in the layers above and below. In meteorites, iridium is one thousand times more concentrated than in ordinary rocks. The idea that the asteroid vaporized ten times its own mass of rock is suggested by the observation that there is only 9 percent as much iridium in the clay as there is in meteorites. In other words, the asteroid was diluted tenfold.

So this great extinction occurred, and a little animal, perhaps a mole or something like that, survived and began evolving into different lines. A little horse preceded the big horses and other animals evolved including cows, mice, rats, and so on. Later, there were primates, rhesus monkeys, and ultimately 4 million years ago, human beings. So perhaps if we do have a nuclear war, there will be some little animal that survives and 64 million years in the future the earth will perhaps again be occupied by intelligent beings. In the case that this happens we can hope they are more intelligent than human beings!

I do not want this to happen. I do not want the second great extinction to occur on earth because of man's inability to control himself.[12]

When called upon, as his time and energy permitted, Pauling would speak out for peace, as he did in 1990 in Los Angeles, at Soka University's Second Pacific Basin Symposium. His topic: "Prospects for Global Environmental Protection and World Peace As We Approach the Twenty-First Century." As usual, he injected optimism into his rather informal talk.

I have great hope. I believe that we are going to be able to solve the great problems that face the world now. The problem of deterioration of the environment, the destruction of the ozone layer, the destruction of the forests, and so on, increasing contamination of the earth by nuclear radioactive substances, the problem of starvation and malnutrition for a major fraction of the world's people. We can't solve some of these problems unless the population explosion problem is also solved. I don't think I need to go into detail about these problems, but I shall repeat that I believe that all of these problems can be solved—if we use good sense in our utilization of the world's resources.[13]

As he grew into old age, Linus Pauling became a sage, an informal philosopher whose statements were reported in different nations of the world. His appearance was engagingly familiar—with a genial smile and sparkling blue eyes, the omnipresent black beret atop a crop of venerable, unruly white hair, a crackling and rather high-pitched voice that he willingly used with the press, whether to speak out on controversial peace issues or to preach the importance of taking vitamins.

Society traditionally honors those tribal elders who have accomplished admirable deeds in their lives. Far more than most mortals, Pauling had achieved much in many different endeavors. He kept adding to the multitude of medals and honorary diplomas that already demonstrated his qualifications as an international wise man.

Years earlier, when delivering his personal credo to the American Humanist Society, Pauling—who had been named Humanist of the Year (1961)— had concluded his talk with an inspiring vision of the part that each individual ideally takes within society. True to his vocation, he couched it in terms that a scientist, particularly a molecular biologist, might especially appreciate. His metaphorical message, then and always, held both warning and hope for future generations of humankind.

MAN has reached his present state through the process of evolution. The last great step in evolution was the mutational process that doubled the size of the brain, about seven hundred thousand years ago; this led to the origin of man. It is this that permits the inheritance of acquired characteristics of a certain sort—of learning, through communication from one human being to another. Thus abilities that have not yet been incorporated into the germ plasm are not lost until their rediscovery by members of following generations, but instead are handed on from person to person, from generation to generation. Man's great powers of thinking, remembering, and communicating are responsible for the evolution of civilization.

Yet a man or woman is not truly an organism, in the sense that a rabbit is, or a lion, or a whale. Instead, he is a part of a greater organism, the whole of mankind, into which he is bound by the means of communication—speaking, writing, telephoning, traveling over long distances, in the way that the cells of a rabbit are interconnected by nerve fibers and hormonal molecular messengers.

This great organism, humankind, is now master of the earth, but not yet master of itself; it is immature, irrational; it does not act for its own good, but instead often for its own harm.

We must now achieve the mutation that will bring sanity to this great organism that is mankind. Must this be a mutation of some of the genes in the pool of human germ plasm? Perhaps such a genetic mutation, providing, for example, extrasensory perception and instantaneous communication among all human beings, would do the job; but I fear that we do not have time for this mutational process to be effective. The human race may cease to exist in a decade.

We must accordingly hope that the mutation can instead be in the nature of the giant organism, humankind, itself—a mutation in the means of communication, in the nerve fibers of the organization, that will transfer to this whole great organism some of the desirable and admirable attributes that are possessed by the units of which it is composed, the individual human beings. The attributes that must be transferred from the units, human beings, to the great organism, humankind, are sanity (reason), and morality (ethical principles).[14]

Quite appropriately, during the very period of this talk, along with his peace-promoting activities Pauling was deeply immersed in research on the biochemistry of the human brain—that control center not just of communication but also, potentially, of reason and ethics (alongside humanitarianism). Though ostensibly he was searching for the causes of mental defects, perhaps too he was unconsciously seeking a way to reprogram physiological processes in the mind to permit people to live peaceably within human society at last.

It is possible, though, that the urgent next step in human evolution that Pauling sought will not come from altered genes or new pharmacological drugs to improve electrochemical function in individual or collective human brains. Instead, it may lie in the worldwide communications revolution already launched through computers and their networks—creations of scientists and engineers that link together individuals, groups, and nations. Toward the close of the twentieth century, the electronic fellowship active on Internet and myriad other digitalized communication channels may begin to achieve Pauling's potent earlier vision of a unified multiorganism called humanity.

IV

Nutritional Medicine

1954–1994

Ten

Mind and Molecules

The brain is the most sensitive of all our organs to variations in its molecular composition, and mental symptoms of avitaminosis often are observed long before any physical symptoms appear.

Many researchers in science and medicine were surprised, even shocked, when Linus Pauling began to give prominent attention to the vitamins and other natural substances that, he claimed, were required by both mind and body to maintain or achieve good health. In 1967 he assigned to this new discipline a name that people thought odd or abstruse— orthomolecular medicine. (Pauling added the Greek word *orthos*, "correct," to the molecular approach taken for years in his medical research.) The last quarter century of Pauling's professional life, it turned out, would be devoted primarily to exploring the promise of orthomolecular medicine and promoting human health.

But perhaps scientists and physicians should have expected something like this from Pauling. There was a logical progression and consistency to this new involvement. The biomedical work Pauling had started doing in the mid-1930s—studying, for example, hemoglobin's oxygen-carrying function, the behavior of antibodies in the immune response, and the molecular structure and function of proteins—amply showed a lengthy and extensive interest in the biochemical aspects of human physiology, body fluids, and cellular metabolism. Increasingly, too, he was combining his social-reformer tendencies with his concern for human-health issues that could be addressed through research.

Pauling's awareness of vitamins was scarcely new. From time to time in his scientific talks and papers he acknowledged the importance of vitamins, minerals, and other nutrients that furnished the biochemical fuels to cells that enable metabolism to go on, assuring growth, move- · ment, and reproduction—life, in other words. By 1938 Pauling was already talking about the challenging research potential in vitamins and other micronutrients, as he did in his dedication speech at the opening of Caltech's new Crellin Laboratory.

ORGANIC chemistry was developed into a great science during the nineteenth century, and it seems probable that all or nearly all its fundamental principles have now been formulated. There is, however, a related field of knowledge of transcendent significance to mankind which has barely begun its development. This field deals with the correlation between chemical structure and physiological activity of those substances, manufactured in the body or ingested in foodstuffs, which are essential for orderly growth and the maintenance of life, as well as of the many substances which are useful in the treatment of disease. These various physiologically active substances are often extremely complex.[1]

The swift expansion of molecular biology into many areas of biomedical research had proceeded from the confirmation of Watson and Crick's structure for DNA proposed in 1953. It enabled researchers to envisage and then initiate all manner of useful developments in medicine through research and genetic engineering. The electron microscope and increasingly sophisticated instrumentation in biochemical crystallography and molecular manipulation permitted ever-closer work in biomedical research.

At Caltech Pauling participated in various DNA-connected molecular-biology investigations. Most notably he worked collaboratively with French molecular biologist Emile Zuckerkandl, who came to Caltech in the fifties to work with him. The two men originated innovative research in what they called paleogenetics—now better known as molecular evolution. Their studies, both theoretical and laboratory based, brought forth the important idea of the molecular clock, which enabled paleontologists and taxonomists to estimate the time when different classifications of plants and animals separated. It also proposed techniques for determining not only how and when mutated or damaged genes in DNA occurred, but also how genetic repair might eventually be done by scientists. (Their collaboration continued after Pauling left Caltech, and

Zuckerkandl later served as director of the Linus Pauling Institute of Science and Medicine for a dozen years.)

In 1959 Pauling remarked:

I am astounded by the rapid progress that has been made during recent years in the understanding of the molecular structure of human beings. Ten years ago I would have said that the discovery of the structure of deoxyribonucleic acid and the formulation of a reasonable molecular mechanism for its self-duplication would be something that our children or grandchildren might achieve, but not we ourselves; yet the Watson-Crick proposal seems to be satisfactory. The exploits of the nuclear physicists during the last few decades show how fast progress can occur, when the time is ripe. I believe that the next fifty years are going to be the golden years for biology and medicine.[2]

In the 1950s and early 1960s, despite his own concurrent energy-demanding involvement in nuclear politics, Pauling was as committed as ever to biomedical research. His customary front-line investigations went on—thanks in good part at Caltech to many coworkers on projects he had initiated and then supervised. Even then, he was making it his business to study health problems, whether or not they were connected with war and peace. For instance, Pauling began applying some of the statistical methods he was using in his antinuclear activism, taking on such health topics as the effects of obesity and smoking on longevity, which he published in Caltech's *Engineering and Science* magazine.

By the mid-fifties, alarmed at the health hazards of fallout, Pauling warned about by-products of nuclear testing that might be present in food. Entering the body, these radioactive forms of elements, such as carbon 14 and strontium 90, could become incorporated in cells. Because strontium was similar to calcium, its dangerous isotope could be absorbed and deposited in the bones, eventually to cause cancer or to harm DNA in the gonads, the reproductive organs of males and females containing the sperm and ova that might produce future progeny. He recommended that people—particularly growing children, whose need for this mineral in bone building is especially great—take calcium supplements made from minerals instead of drinking milk or eating milk products possibly contaminated with strontium 90 coming through the food chain from plants and animals that had absorbed the radioactive substance. The health-protecting need to avoid, restrict, or find an analog for an otherwise important nutrient in the diet was already familiar

to him personally because of his own earlier special diet restricting animal protein, due to glomerular nephritis.

Pauling found it outrageous and unjust that people now had to put up
with nuclear fallout created by nations' extreme economic and military
rivalries. There were enough infectious and genetic diseases and other
health problems already, such as famine and malnutrition, that caused
great suffering in the world, without introducing more. In the early
1960s Pauling joined those who were rapidly becoming alarmed about
the degradation of the environment, particularly the prevalence of many
toxins—poisons and carcinogens—that originated in industry, agricultural usage, and food processing and preservation. Like radioactive fallout from nuclear testing, these toxins contaminated the air, soil, and
water. And they affected humans and all other forms of life.

WE have been living for some decades in an environment that contains
ever-increasing amounts and varieties of molecules that were not present during the period of evolution of the human race. These are the
molecules of substances that do not occur in nature and have been only
recently synthesized by chemists for use as insecticides, as additives to
foods, as additives to gasoline, as drugs, and for other purposes. Some
of these substances are toxic to human beings.[3]

Pauling, who had originated the concept of molecular disease when discovering the cause of sickle-cell anemia, knew that abnormalities were
caused by random mutations in genes. Toxic chemicals and high-energy
radiation, he said, could adversely alter DNA in individuals exposed to
them. Mutations cause abnormal cell growth such as cancer, disable
critical physiological systems such as enzyme production, or produce
physical or mental defects—creating molecular diseases—in offspring
produced by the affected male or female parent. Thus Pauling joined
other scientists and public-health advocates in advocating rational ways
to halt unmindful births so as to eliminate defective genes and promote
superior ones in human reproduction. He used logic, statistics, and compassion—in the latter case, an effort to minimize human suffering—in
his forays into the still-controversial topic of eugenics, in both its positive and negative forms. That is, he encouraged the reproduction of
people with favorable genetically transmitted characteristics, such as
superior intelligence and good mental and physical health, and coun-
seled against transmitting known molecular diseases through the "marriage" (Pauling was old-fashioned in his views of procreation) of
heterozygotes—people who are carriers of defective genes without hav-

ing the disease itself (such as the one that causes sickle-cell anemia). The progress in prenatal testing and advanced in vitro fertilization techniques, the legalization of abortion, and the promise of corrective treatment through genetic engineering—all, theoretically at least, offer ways to implement the eugenic programs that Pauling espoused in a number of papers and talks to physicians and biomedical researchers, especially during the 1950s and 1960s.

By then, too, he acknowledged the importance of thousands of biochemical substances in body fluids essential to health. These were the hormones that functioned as messengers and the enzymes needed as catalysts to carry out the basic cell and body processes by facilitating metabolic reactions and converting one substance into another, whether to serve as a nutrient, neutralize toxins, or be excreted. He increasingly realized that many molecular diseases, identified or yet to be identified, were connected with serological, immunological, or enzyme malfunctions that affected basic metabolism, cell growth and replacement, and homeostasis. They might therefore be corrected, alleviated, or even prevented through nutritional medicine—through the avoidance or supplementation of particular dietary substances that were chemicals natural to the body. (For example, it was found that maintaining a high level of urea, or uric acid, in the bloodstream often averted or eased patients' crises with sickle-cell anemia.)

In that period of time—the fifties and early sixties—Pauling took particular interest in enzymes.

THE enzymes constitute a class of protein molecules of extraordinary significance. Just as deoxyribonucleic acid may be said to be the substance that we need to understand if we are to know in its atomic detail the mechanism by means of which organisms reproduce themselves, so are the enzymes the molecules that we must understand to know how organisms carry on their metabolic activities.[4]

With his new interest in genetic and molecular diseases, Pauling was zeroing in on the role of ill health, including mental disorders, that various enzymatic malfunctions might take.

THE problem of enzymes encompasses essentially the whole of biology. When we understand enzymes—their structure, the mechanism of their synthesis, the mechanism of their action—we shall understand life, except for those aspects of life that involve mental processes; and I have no doubt that enzymes are important for these too.

Enzymes do an extraordinary job—that of causing chemical reactions to take place in the body, at body temperature, which without enzymes can be made to take place only under much different conditions or with great difficulty. . . .

Perhaps [a genetic] abnormality is such as to lead to a complete interruption of the process of manufacturing [an] enzyme, or perhaps an abnormal molecule is manufactured. . . . A score of diseases have so far been recognized as enzyme diseases, presumably resulting from the manufacture of abnormal enzymes in place of the active enzyme molecules. I think that it is not unlikely that there are hundreds or thousands of such diseases.

I foresee the day when many of these diseases will be treated by the use of artificial enzymes.[5]

Pauling was developing a concerted approach to human health that was based on the biochemistry involved in the interaction of genetics, environment, age, and diet. In a 1958 article in Pfizer Laboratories' magazine *Spectrum,* after discussing sickle-cell anemia, Pauling explained some of the elements relating to enzyme dysfunction that would emerge later in his postulation of orthomolecular medicine.

As to other molecular diseases, we should probably have to include those hereditary disturbances in which a gene is so abnormal that its product is entirely absent. It is likely that in agammaglobulinemia there is a gene defect that leads to the production of either very little or no gamma globulin. Perhaps normal gamma globulin *is* produced, but there is too little to do any good. Perhaps, instead, an *abnormal* globulin is being manufactured and deposited somewhere in the body, although as yet such a thing has not been demonstrated. Similar diseases are acatalasia, phenylketonuria, and galactosemia; probably in all these an enzyme is effectively missing, and a metabolic product accumulates to the point where it becomes toxic. In galactosemia, the enzyme galactose-1-phosphate-transferase is not present, so that galactose accumulates and produces mental deficiency and physical illness—unless it can be excluded from the infant's diet in the first place, by removing milk from the diet.

What must be learned is whether the defective gene produces a molecule that fails to work, or a molecule with undesirable properties (as in sickling), or no molecule at all. When, in the general sense, all these alternatives are considered, a disease such as atherosclerosis, with its prominent familial character, is probably also a molecular abnormality

—one in which the subject is genetically left unable to withstand environmental stress due to the nature of his occupation or due to certain foods which he may take to excess. If his genes had been different, his molecules would have been different, and theoretically he could have withstood the adverse condition without developing disease.

Many of the possible etiologic mechanisms of cancer reduce to the problem of a molecular disease also. It is likely that this is true of all diseases having a clearly hereditary trend. Included among them is the very great problem of mental deficiency, for, as you may know, the curve for the distribution of the population by intelligence does not follow entirely the bell-shaped curve or Gaussian error function; rather, there is a rise at the low end which can be accounted for by mental deficiency due to such accidents as the inheritance of homozygous genes for galactosemia or phenylketonuria.

I have been hesitant—perhaps not hesitant enough—to say much about mental disease. I think that mental deficiency is often the result of a qualitative abnormality, and mental illness the result of a quantitative abnormality. That is, mental deficiency can result from complete absence of certain enzymes required for bodily function, while mental disease can be due to the presence of only half as much of certain enzymes as may be normally required, so that the person is less well able to withstand environmental difficulties. The vital chemical constitution of a person is, I believe, one factor; environmental burdens are another.[6]

Pauling was also clearly interested in discovering ways to slow down the aging process, which he noticed in himself, as he candidly revealed during this lecture. At the same time, he was evaluating his decade-long involvement in psychiatric research, which had begun in 1954.

I'M an optimist. I hope to live a long time. I don't smoke cigarettes, never have smoked cigarettes, and so I am some seventeen years younger physiologically than a person who has just had his sixty-fifth birthday but has been smoking two packs of cigarettes a day since he was twenty. So I expect to live a long time. I don't live in New York, and don't have to breathe its atmosphere except when I am forced to come here. . . .

I like human beings. I like to think about the possibilities of decreasing the amount of their suffering. As they grow older, of course, the incidence of disease becomes greater. With me, the incidence of disease at age sixty-five is about the same as for a cigarette smoker at age forty-eight. But I still feel it; my knees begin to get shaky after I have stood as long as I have now. I enjoy life. I look around for samples of hemoglobin, and in *other* respects I enjoy life. I like to think that we are going to be able to improve the feeling of well-being of older people through the discoveries about—let's say about the nature of the brain.

Why is it that older people begin to suffer from loss of memory, loss of mental activity? I don't have anything special to say about this, but I am confident that chemotherapeutic means of preventing this deterioration in mental activity in older people can be developed and *will* be developed.

When I remember that 10 percent of the American people spend some part of their lives in a mental hospital, that half of the hospital beds in the country are occupied by mental patients, then I like to think that I worked for some years in the field of the chemical basis of mental disease and got some younger people started off in this field, and that many other people are attacking this very difficult problem.

I like to believe, as I do believe, that it will be possible to get an understanding of the molecular and genetic basis of mental disease and also, of course, better understanding of the influence of environmental factors in producing mental diseases. I believe that it will be possible in the course of time to carry out significant differential diagnoses of mental disease; that we shall, in the course of time, be able to develop therapeutic methods that will lead to the effective control of a very large part of mental disease.

And, of course, I believe also that when we eliminate war from the world, and many of the injustices that exist at the present time, the amount of mental disease in its response to *environmental stress will decrease.*[7]

Pauling's interest in disturbed mental function inevitably focused on physiology instead of standard psychotherapy. In his researches on the chemical bond in molecules, the attraction of oxygen to hemoglobin in blood, and the forces that joined antigen to antibody in the immune response, Pauling often dealt with ions. Many atoms have minute electrical charges, involving the presence of extra electrons or the absence of them, which make them negative or positive, which thereby influences the types of bonds formed with other atoms in molecules and compounds. He knew that the brain's nerve cells, or neurons, which receive and transmit or store information and instructions, have an essentially electrochemical circuitry. Impulses going back and forth would move through nerves rather like electrons through highly conductive copper threads in electrical wiring or telephone lines. Some points would send and others would receive messages, with chemicals—such as enzymes —influencing the necessary activation or inhibition of impulses along neurons.

Certainly, Pauling reasoned, adequate brain function—viewed from what might be considered the specialty of brain-fluid chemistry—depended upon receiving through blood circulation constant supplies of oxygen from the air and nutrients from food, both needed for metabolism that produces energy. Blood chemicals would also require having proper electrolytic balances for normal consciousness and mental function in brain tissue, and for nervous-system cells generally. Mentality, then, was really as much a proper province for the biochemist as it was for the psychologist and psychiatrist.

Pauling's knowledge of and vocabulary for mental function and brain physiology possibly originated in 1948, when his childhood friend Lloyd Alexander Jeffress, who had become a research professor in psychology, spent a sabbatical year at Caltech, thanks to a generous bequest to the institute designated for psychological research. Pauling had made this arrangement by recommending that Jeffress, interested in both physics and psychology, be asked to put together a major conference, "Cerebral Mechanisms in Behavior"—which Pauling, back from his own stay in England, was fortunately able to attend. (This was the start of Caltech's notable entry into the realm of physical psychology, headed by Roger Sperry.)

Pauling must have been giving considerable thought to the human brain and its physiological and psychological disorders. For in 1953, when his principal colleague on the Caltech hemoglobinopathy studies, Dr. Harvey Itano, accepted a position at NIH, Pauling was ready anyway to move into another research area. He was well aware of a built-in cycle in sustaining intensely focused interest; he tended to work on major projects lasting about five years. He began considering where his research skills might best be utilized.

IN 1953 I decided that I would do something other than study the hereditary hemolytic anemias. I thought that since nobody is working in this new field of molecular diseases, I might as well work on an important disease. What are the important diseases? Cancer and mental illness. But everybody worked on cancer then and nobody worked on mental disease, so I decided to work on mental disease.[8]

Naturally Pauling expected to find that many mental disorders were caused not by psychological and environmental factors—such as repressed emotions in early childhood development and pathological family relationships, as the psychoanalysts insisted—but by molecular diseases, possibly some deficiency or excess of a particular enzyme or other brain-affecting body chemical due to a genetic defect. (Eventually he'd get into cancer research too.)

In 1954 Pauling applied for and got major research funding from the Ford Foundation, later supplemented by NIH and NIMH grants, to determine what genetic and biochemical disorders might be involved in mental retardation and schizophrenia—with the possibility then of finding ways in which they could be prevented or ameliorated. When mongolism (Down's syndrome) was determined to be caused by the abnormal presence of an extra chromosome, this form of retardation no longer qualified for Pauling's attention or renewed funding. He focused now on retardation and acute mental illness caused by biochemical dysfunctions, as with the absence or overabundance of critical enzymes. His extensive prior work with the hemolytic anemias gave him confidence in his subsequent work in seeking genetic bases for particular mental disorders and deficiencies.

In 1956 the *Saturday Review* asked a group of notable scientists to identify "new areas for fruitful research" for an article called "The Research Frontier." Pauling was quick to respond.

MENTAL disease, including mental deficiency, is one of the greatest medical problems. It is likely that most cases of mental deficiency are caused by a molecular abnormality of some sort. It is known that one kind of mental deficiency involves the amino acid phenylalanine, but the way in which phenylalanine or a substance formed from it operates to produce the mental manifestations is not known. Further biochemical research on mental deficiency, with special attention to molecular abnormalities, should lead to significant progress.[9]

Not long afterward, Pauling would be able to explain the cause and control of this particular molecular disease called phenylketonuria (usually known as PKU). In 1962 he was asked to contribute to the four-day proceedings of the Second International Conference of the Manfred Sakel Foundation, which was held in autumn of 1962 at the New York Academy of Medicine. He delivered the paper "Biological Treatment of Mental Illness." Characteristic of the time (the test-ban treaty and the Nobel Peace Prize were yet to come), Pauling's introductory remarks referred to technological advances, including nuclear weapons, and irrational behavior by national governments. Then he got to the main topic.

I N some parts of the world starvation, malnutrition, and the onslaughts of infectious diseases are the principal causes of human suffering; but everywhere mental disease is an important cause. In the United States half of the hospital beds are occupied by mental patients.

Several years ago I became interested in the question of the chemical basis of mental disease. As a result I have had contact with many psychiatrists. I am grateful to them, and I hope that they will not have to work much longer under the terrible handicap of extreme ignorance about the diseases that they are treating and about the mechanisms of their methods of treatment. I have been encouraged by information presented at the Manfred Sakel Conference to believe that the time will come, perhaps in ten or twenty years, when, if enough effort is made, there will be obtained some significant understanding of the nature of the groups of diseases that we classify as schizophrenia and of other mental diseases, comparable to that which now exists for a few diseases that are called molecular diseases.

I believe that most mental diseases are molecular diseases, the result of a biochemical abnormality in the human body. I think that the mind is a manifestation of the structure of the brain, that it is an electrical oscillation in the brain supported by the material structure of the brain; I think that the mind can be made abnormal by an abnormality in the chemical structure of the brain itself, usually hereditary in character, but sometimes caused by an abnormality in the environment. Pellagra is an example of a disease with mental manifestations, which has been largely conquered because of the discovery of the molecular abnormality that produces it. It is a deficiency disease, due to the lack of a vitamin, nicotinic acid. The discovery of its chemical cause has led to the solution of the pellagra problem in many parts of the world.

Considerable progress is now being made in the understanding of molecular diseases that cause mental deficiency. An example is the disease phenylketonuria. It was identified thirty years ago by Dr. Foelling in Oslo, Norway, who studied some mentally deficient children who had an odd smell. He found that there were some unusual substances present in the urine of these children; further investigation by him and other scientists then led to the discovery that these children lacked an enzyme in the liver that catalyzes the oxidation of one amino acid, phenylalanine, to another amino acid, tyrosine. In consequence, the concentration of phenylalanine in the blood and cerebrospinal fluid increases to such a level as to interfere with the development and function of the brain, leading to mental deficiency. The various proteins that we eat contain about twenty different amino acids. The amount of phenylalanine in the diet is somewhat greater than is needed, and there may be a deficiency in tyrosine. Normal human beings have an enzyme in the liver that catalyzes the oxidation of the excess phenylalanine to tyrosine. This enzyme is manufactured by genes—most people have two genes of this sort, which have been inherited from their parents. One person in eighty has only one gene that manufactures the enzyme, plus another gene, a mutated one, that does not manufacture the enzyme. Such a person has only half the amount of the enzyme that a normal person has, which is, however, enough to keep him in good health.

If two of the people marry one another, there occurs the great lottery: the child inherits only one of the two genes from the father and one of the two from the mother. On the average, a quarter of the children from such a marriage thus inherit the abnormal gene of the father and also the abnormal gene of the mother, neither of which will manufacture the enzyme. In consequence, this child has the disease phenylketonuria.

When the knowledge of the chemical basis of phenylketonuria was obtained, it was found that the disease could be recognized by a chemical test a few weeks after birth, and that the child, if fed on a diet of hydrolyzed protein from which most of the phenylalanine had been removed, would develop in a nearly normal manner.

This disease, which in the past has been responsible for one percent of the institutionalized mentally deficient individuals in the United States, may now be brought under control because of the discovery of its molecular nature.

This example indicates what we might expect to find in schizophrenia —not that there is a gene for schizophrenia, but rather that there are many genes involved, such that any of them, when inherited in double dose, some combinations of several of these genes with one another may produce a quantitative genetic abnormality that leads to mental manifestations classified as schizophrenia. The evidence that was presented to the symposium about the presence of biochemical abnormalities in the blood of schizophrenics is significant in this respect.

If schizophrenia is a complex of diseases, rather than a single disease, it will be difficult to find a treatment. Nevertheless, I believe that a tremendous amount of progress can be made in controlling schizophrenia, and in decreasing the amount of human suffering that it causes, provided that a vigorous effort is made to carry out research on its nature.

We may envisage ways for treating diseases that have not yet been brought into practice. For example, with the progress in our knowledge about the nature of enzymes and about the role of enzymes in relation to disease, especially enzyme deficiencies, it may before long be possible to synthesize artificial enzymes that can be used in a substitution therapy that will permit the patient to lead a satisfactory life.

We have now to recognize the possibility that nucleic acid may be introduced into the cells of a patient, to correct the abnormalities that otherwise would lead to mental disease. There is also the possibility that an understanding of mental disease will be obtained such that drugs of definitely predicted structure can be synthesized that will control the diseases in a specific way.

However, all of these would be palliative measures, which would not constitute a solution of the problem. I am concerned about the degeneration of the pool of human germ plasm, resulting from damage to the genes by natural mutagenic agents, natural radioactivity, and those chemicals and artificial high-energy radiation that constitute a part of our modern life.

There is a natural process of purification of the pool of human germ plasm. As abnormal genes, such as those that produce phenylketonuria, continue to be introduced into the pool of human germ plasm through new mutations, the defective genes are also removed by the death without progeny of those children who have inherited the abnormal recessive gene in double dose. This process of purification involves a large amount of human suffering.

As our knowledge about the genetic character of human beings increases, it may become possible to identify the carriers of the recessive genes for many serious diseases. It would then be possible for these carriers to marry normal individuals, rather than other carriers of the same defective genes, and then to have a somewhat smaller number of children than normal, one or two children, rather than three or four. In this way the defective genes would be removed from the pool of human germ plasm, without the birth, suffering, and death of defective children. As the years go by, there may be achieved by this method a significant decrease in the incidence of mental disease, as well as the diseases with somatic manifestations.

I believe that we shall move into a better world in the future, with the scientists, collaborating with the psychiatrists, in doing their part to contribute to the control of mental disease and the decrease in the amount of human suffering. We may have hope that the increase in knowledge about the molecular basis of mental disease will in the course of time lead to a significant decrease in the amount of human suffering in the world.[10]

In an address given to a meeting of American psychiatrists in 1959, Pauling discussed his explorations of connections between molecular disease and particular mental disorders. Given the time period in which this and other talks and writings on the subject were done, it is scarcely surprising that Pauling pointed out that new "mutagenic agents," including nuclear-weapons testing and environmental stressors, would exacerbate the problem of genetic diseases, including those affecting the mind—which had become his main area of research.

IT is only ten years since the molecular basis of sickle-cell anemia and of the other hemoglobinemias was discovered, and the details of the difference in molecular structure of the hemoglobins responsible for the diseases have not yet been worked out. We may hope that as more and more information is gathered about the molecular nature of these diseases it will, in the course of time, be possible to develop on the basis of this information an effective therapy for them, and thus to decrease the amount of human suffering that these mutant genes cause. . . .

The exposure of the gonads of Americans to background radiation, from cosmic rays and natural radioactivity, amounts to about three roentgens during the first thirty years of life. Medical X rays provide approximately the same exposure. The estimate that I have made of the exposure due to radioactive fallout, for nuclear explosions carried out at the average rate of the past few years, is about 0.3 roentgen in thirty years.

There is nothing that we can do about the damaging of the pool of human germ plasm by cosmic rays and natural radioactivity. It is, of course, possible by good medical practice to decrease somewhat the exposure of the gonads to medical X rays, as by shielding the gonads whenever an exposure of some other part of the body is being made. Another effective way is through the use of X-ray films of increased sensitivity, and through the limitation of roentgenographic examination to cases when there is significant medical justification.

The damage done by radioactive fallout can be limited through the cessation of the tests of nuclear weapons. It is worth while to attempt to make an estimate of the number of human beings who might be kept sane by the act of stopping the testing of nuclear weapons.

The effects that we shall discuss are genetic effects that would appear in the population in the United States during the next few generations. I shall assume that the average population of the United States during the next few generations will be 200 million. Of this total, 10 per cent, 20 million, may be expected to spend some time during their lives in a mental hospital. I assume that there is a strong hereditary factor involved for 80 per cent of them; that is, for 16 million. If the testing of nuclear weapons were to be continued at the recent annual rate for a period long enough to expose the germ plasm of the entire population to the fallout dose, an estimated number of mutations corresponding to an additional 160,000 cases of mental illness would be produced in the population of 200 million. The tests that have already been carried out can be similarly estimated to cause mutations that will result in serious mental disease in twenty-three-thousand people in the United States. I think that it can be a source of satisfaction to us that at the present time no nation is continuing its tests of nuclear weapons, and that the Geneva Conference on an International Agreement for the Cessation of Nuclear Explosions is continuing, and may soon lead to the formulation of an effective agreement.[11]

Pauling's involvement with brain functions inevitably caused him to spend more time considering how the mind works—and specifically in scientists. He had always been interested in the process of discovery, and in 1955 he prepared a rather philosophical talk, "The Genesis of Ideas," to be given at a meeting of the American Association for the Advancement of Science (AAAS). It was published some years later, after he slightly revamped the lecture for presentation at a conference of psychiatrists in 1961.

HAVING ideas is a part of thinking. Some people have said that thinking is the process of solving problems. Professor John Cohen said some years ago in *Nature* that thinking is much broader than this in scope: that in much of our thinking we are groping to find out *what* needs to be done rather than *how* it needs to be done. Some of the most significant new ideas in science involve the recognition of new problems.

Many scientists have been interested in the question of the way in which scientific discoveries are made. A popular idea is that scientists apply their powerful intellects in the straightforward, logical induction of new general principles from known facts and the logical deduction of previously unrecognized conclusions from known principles. This method is, of course, sometimes used; but much advance in knowledge results from mental processes of another sort—in large part unconscious processes. Henry Poincaré in his essay on mathematical creation said that knowledge of mathematics and of the rules of logic is not enough to make a man a creative mathematician; he must also be gifted with an intuition that permits him to select from among the infinite number of combinations of mathematical entities already known, most of them absolutely without interest, those combinations that will lead to useful and interesting results. . . .

From my own experience I have come to the conclusion that one way for me to have a new idea is to set my unconscious to work on a problem. This is probably what the Persian philosopher Avicenna, a thousand years ago, also did when he was unable to solve a problem. He would go to the mosque and pray for his understanding to be opened and his difficulties to be smoothed away; he probably had fixed the problem in his mind before going to the mosque, and his nature was such that his unconscious could then set to work on it.

I doubt that the unconscious can be directed to work on a problem. But the problem can be suggested to it, and if it is interested in it something may result.

C. G. Jung has said that art is a kind of innate drive that seizes a human being and makes him its instrument. A creative scientist is an artist—an artist whose ideas are in the field of science. . . .

Herbert Spencer has written that it was never his way to set himself a problem and puzzle out an answer; that instead the conclusions at which he had from time to time arrived had been arrived at unawares, each as the ultimate outcome of a body of thoughts that slowly grew from a germ.

I once heard a commencement speaker say that there is no place in the world for the man who works just to satisfy his own curiosity, that instead everybody should work for the solution of problems that will benefit the world. He was an engineer, primarily interested in the application of knowledge to the solution of practical problems. But he had forgotten that the knowledge that he wishes to apply must first be obtained, and that to obtain it is not easy. The knowledge that we have about the world, and that is applied, for the benefit of mankind, to the problems of technology and medicine, has in the main been obtained by men who were satisfying their curiosity about the nature of the world, by pure scientists and mathematicians.

The progress in the physical sciences and in technology that has taken place during recent decades has been based in large part on the developments in mathematics of the last two hundred years. Most of the mathematicians whose work has turned out to be fundamental to practical progress did not make their mathematical discoveries for the use of the engineer, but rather to satisfy their own curiosity. Often a field of mathematics is not found to be valuable in technology for a long time— it is rarely that a mathematician sees his discoveries put to practical use during his lifetime.

I could mention many examples of important discoveries in the field of physics and chemistry that have resulted from the effort of a scientist to satisfy his curiosity. It is, of course, possible to be curious about trivial matters. . . . Albert Tyler has told me that Thomas Hunt Morgan once said to him that a little idea is enough to permit a scientist to get started. He told Tyler that even a simple experiment, when carried out, may suggest another one, and that another one, until, as the result of a succession of experiments and ideas, some significant discovery is made. As a geneticist, he might have said that ideas breed ideas. . . .

We need to study the general problem of the genesis of ideas. I think that it is not unlikely that young men can be given some training in having ideas—would not instruction in the art of having ideas about good experiments be just as valuable to the young experimental scientist as instruction in laboratory technique? I doubt, however, whether we know enough at the present time about the genesis of ideas to plan a course of instruction.

My own experience, which I may illustrate in my talk by some examples, has suggested to me that it is possible to train the unconscious to help in the discovery of new ideas. I reached the conclusion some years ago that I had been making use of my unconscious in a well-defined way. I had developed the habit of thinking about certain scientific problems as I lay in bed, waiting to go to sleep. Sometimes I would think about the same problem for several nights in succession, while I was reading or making calculations about the problem during the day. Then I would stop working on the problem, and stop thinking about it in the period before going to sleep. Some weeks or months might go by, and then, suddenly, an idea that represented a solution to the problem or the germ of a solution to the problem would burst into my consciousness.

I think that after this training the subconscious examined many ideas that entered my mind, and rejected those that had no interest in relation to the problem. Finally, after tens or hundreds of thousands of ideas had been examined in this way and rejected, another idea came along that was recognized by the unconscious as having some significant relation to the problem, and this idea and its relation to the problem were brought into the consciousness.

As the world becomes more and more complex and the problems that remain to be solved become more and more difficult, it becomes necessary that we increase our efforts to solve them. A thorough study of the general problem of the genesis of ideas and the nature of creativity may well be of great value to the world.[12]

In the newer version of "The Genesis of Ideas," Pauling told the story of one of his breakthrough insights into the creative process, which he

thought exemplified how a scientific insight or discovery might originate, even after lying dormant for some years. He recalled this sequence in his single-chapter unpublished memoir.

IN preparing for the Third World Congress of Psychiatry, to be held in Montreal, Canada, in 1961, the organizers of the congress decided that they would arrange for a special symposium to be held, dealing with the origin of ideas. They invited three speakers, who were, I suppose, people with a worldwide reputation for having ideas. . . .

In my talk to the psychiatrists I gave an example of a problem that I had worked on for seven years before having the idea that provided a solution to the problem. I said that I had served for about a decade as a member of the Scientific Advisory Board of Massachusetts General Hospital in Boston. At one annual meeting of the board, in 1952, the hospital had arranged that a talk about anesthesiology be given to members of the board by the professor of anesthesiology in Harvard University, Dr. Henry K. Beecher. In the course of his talk Dr. Beecher mentioned that the noble gas xenon was a good anesthetic agent, that two operations had been carried out on human beings using it as the general anesthetic. Xenon is too expensive to be used as an anesthetic in operations, although it may be superior to the substances that are generally used. Up to that time, I had not thought about the problem of how an anesthetic agent works. I think that I had thought that diethyl ether, chloroform, nitrous oxide, halothane, and other substances probably reacted with constituents in the brain to cause anesthesia. However, here was a fact that puzzled me, because xenon could not possibly react with substances in the brain to form chemical compounds. It is called a noble gas or inert gas because of its essentially complete lack of chemical reactivity.

My son Linus Jr. was then in his last year as a medical student in Harvard Medical School, and I was staying with him and his wife. When he picked me up after the lecture, to take me to his home, I told him about this fact, that xenon served as an anesthetic agent, and asked him if he understood how it could have this property. His answer was that he had no idea about how xenon might work or how any other general anesthetic works in the human body or in animals to produce unconsciousness. This fact, that the unreactive gas xenon could have an important physiological property, of causing human beings to become unconscious, puzzled me, and for several nights I continued to think about it, after I had gone to bed and before I had gone to sleep. I believe that during these times, when I was puzzling over this fact, I trained my unconscious mind to continue to remember that this problem existed.

Perhaps my unconscious mind considered every new piece of information that entered my consciousness, to see whether or not it might have some connection with the anesthetic activity of xenon.

Seven years went by. One morning in my laboratory in Caltech, as I was reading my mail, I came across a paper that had been sent to me by a friend of mine, an X-ray crystallographer. As I was reading this paper, I suddenly became wide awake and said to myself, "I understand anesthesia!" I had the idea that consciousness and short-term memory consist of electric oscillations in the brain, with permanent memory involving the laying down of molecules in the brain in such a way as to form a pattern that can interact with the electric oscillations. These electric oscillations, I thought, interact with electrically charged atoms in the brain. The xenon or other anesthetic agent, I thought, can cause water in the brain to congeal into very small crystals, stabilized by the anesthetic agent and entrapping the electrically charged atoms in such a way as to interfere with the processes that we describe as consciousness. I was excited by this idea, as I always am when I think that I have discovered the solution to some problem that had not yet been solved by me or by anyone else.

In fact, there had existed for sixty years one theory of anesthesia. This theory had been formulated independently by two investigators, a man named Meyer in Germany and a man named Overton in England. Meyer and Overton knew, of course, that the brain contains a large amount of fat. There is a layer of fat around every nerve. It is believed that the fat may serve as an electric insulator, so that electric signals can be sent along the nerve and from one part to another in the brain, without leaking into the surrounding tissues. Meyer and Overton had noticed that the anesthetic agents are all soluble in fat (olive oil was the fat usually used in the experimental studies) and that the more powerful anesthetic agents had a greater solubility in fat than the less powerful ones. Their theory of anesthesia was that when an anesthetic agent dissolved in fat and reached a certain critical concentration it interfered with the functioning of the brain in some way not understood, such as to cause unconsciousness to supervene. I was able to point out that anesthetic agents can also be made to crystallize with water, forming crystalline hydrates, and that these crystalline hydrates are formed more rapidly by the powerful anesthetic agents than by the less powerful ones, so that in this respect my hydrate-microcrystal theory of anesthesia was just as good as the Meyer-Overton fat-solubility theory. Moreover, I felt that I had a rather detailed explanation of the mechanism of anesthesia, the idea being that when the hydrate microcrystals form they trap electrically charged atoms in the brain and thus interfere with the electric oscillations that constitute consciousness, and hence give rise to unconsciousness.

At the present time there is still some doubt about the mechanism of action of anesthetic agents. It is my belief, and I think also the belief of most anesthesiologists who think about this matter, that the whole story has not yet been told. Also, however, there seems to be quite good evidence that the hydrate-microcrystal idea may well be an important part of the accepted theory of general anesthesia when this theory is finally formulated. In fact, in the spring of 1991 the American Society of Anesthesiology arranged for me to make a videotape about the discovery of the hydrate-microcrystal theory of anesthesia, to be placed in its archives.[13]

Pauling's paper "A Molecular Theory of General Anesthesia" was published in *Science* in June 1961, after he had thoroughly studied the research literature and conducted laboratory work with his colleague Richard Marsh. This was almost a decade after he had asked the initial question about how xenon might work. By the time of the sudden revelation, he had already been studying brain chemistry for five years.

The introductory paragraph revealed the basis of Pauling's interest in mental functions. And here he was telling the world of science that there was a great deal of work to be done now on the physiology and biochemistry of the brain.

DURING the last twenty years much progress has been made in the determination of the molecular structure of living organisms and the understanding of biological phenomena in terms of the structure of molecules and their interaction with one another. The progress that has been made in the field of molecular biology during this period has related in the main to somatic and genetic aspects of physiology, rather than to psychic. We may now have reached the time when a successful molecular attack on psychobiology, including the nature of encephalonic mechanisms, consciousness, memory, narcosis, sedation, and similar phenomena, can be initiated. . . .

It is likely that consciousness and ephemeral memory (reverberatory memory) involve electric oscillations in the brain, and that permanent memory involves a material pattern in the brain, in part inherited by the organism (instinct) and in part transferred to the material brain from the electric pattern of the ephemeral memory. The detailed natures of the electric oscillations constituting consciousness and ephemeral memory, of the molecular patterns constituting permanent memory, and of the mechanism of their interaction are not known.[14]

It was Linus Pauling's involvement in brain research that led him to invent the concept of orthomolecular medicine. In later years he told different versions of the story, originating in what he called orthomolecular psychiatry. Here is one of them.

WHILE we were working on schizophrenia, I learned about two men working in Canada, Abram Hoffer and Humphry Osmond, who in 1952 had observed that some patients with schizophrenia improved greatly when they were given tremendously large doses of niacin or niacinamide, two forms of the pellagra-preventing factor (now classed as B vitamins) that had been discovered back in 1937. Their schizophrenic patients were given doses of 17 grams a day. The amount recommended by the Food and Nutrition Board (the RDA, or recommended dietary allowance) is 17 mg a day; that is, they were giving these patients a thousand times the recommended amount of this vitamin. They and other investigators later found that similarly large doses of vitamin C were also very beneficial to patients with schizophrenia and other mental illnesses.

There is much evidence that the functioning of the brain is determined by the molecular structure of the brain. I have mentioned the anesthetic agents, which cause unconsciousness. Substances such as LSD, marijuana, and hashish affect the functioning of the brain and may cause permanent damage.

In addition to these substances, there are many substances that are required for the proper functioning of the brain. Fifty years ago there were in the world hundreds of thousands of people who were psychotic because of pellagra. When these people were given a few milligrams of niacin per day (nicotinic acid, a form of vitamin B_3), they recovered from both the mental manifestations and the physical manifestations of the disease. Mental manifestations are also associated with scurvy, which results from a deficiency in vitamin C, and also from a deficiency in several other vitamins, of which I may mention pyridoxine, riboflavin, cyanocobalamin (B_{12}), folic acid, and pantothenic acid.

Patients with pernicious anemia or with a deficiency of vitamin B_{12} for some other reason are observed often to suffer from psychosis for several years before any of the physical manifestations of the avitaminosis manifest themselves. In one study in Norway, 15 percent of the mental patients admitted to a mental hospital over a period of one year were found to have a pathologically low level of vitamin B_{12} in the blood, less than 150 picograms per milliliter. The investigators stated that these patients were all given injections of the vitamin and that this led to improvement in their mental health. Similar evidence for some other vitamins, indicating improvement in mental health on ingestion of large amounts of the vitamins, exists in the literature.

There is no doubt that these substances, which are normally present in the brain and are required for proper functioning of the brain, have an influence on the mental state of a person, and there is the possibility that varying their concentrations can be of value in controlling mental disease in some patients. Many of these substances are nontoxic; ascorbic acid, niacin, and niacinamide are less toxic than sodium chloride [salt] and probably no more toxic than ordinary sugar. I believe that if varying the concentrations of these substances can be effective in controlling mental illness, this sort of therapy is far more desirable than the treatment of the patient with powerful drugs, such as chlorpromazine, or with electric shock therapy. I believe that physicians and medical research men should make a much more thorough study of this possibility than has been made in the past.[15]

Pauling being Pauling, he could be expected to say more than this, with pertinency to prevalent social problems.

THERE is another aspect of nutrition that I may mention. About 20 percent of the American people, 40 million of them, suffer from starvation or malnutrition. In some communities in the United States 90 percent of the children come to school without having had any breakfast. Hundreds of thousands of children are born each year in the United States with significantly decreased intelligence because of damage to the brain in the fetal state and in infancy and childhood because of malnutrition of the mother and the child. We, in this affluent country, have the duty to rectify this terrible condition.

About 60 percent of the people in the world, numbering about 2 billion, the poor people of the world, live on 10 percent of the world's income. Their average income is $100 per year. Their total income, $200 billion per year, is equal to the amount of money wasted by the nations of the world on militarism. I mention this great injustice in order to point out that there are many problems that remain to be solved. I believe that, through the discoveries that have been made by scientists and that will be made by scientists, and through the development of a general feeling of moral responsibility for humanity as a whole, it will be possible for the world to be made a much better place during the coming century than it is now.[16]

Elsewhere this outspoken public-health advocate delivered a provocative lecture in which he tried to convert physicians in training to social activism and possibly even a modern social gospel.

STARVATION and malnutrition present a medical problem as well as a social problem. And crime is also a medical problem. I read last week that a man who had killed a woman was acquitted in Melbourne, Australia, because microscopic studies of his cells showed that he had forty-seven chromosomes, an extra Y chromosome. It has been found, in several countries, that 1 to 5 percent of male criminals over six feet tall have this abnormal chromosomal constitution, an extra Y chromosome. They are criminals because—with little doubt—of this abnormality. But aren't most criminals criminals because of either a genetic abnormality or the faults of society?

In the United States we are following the wrong course. I think that the world as a whole is improving, but that we have been going in the wrong direction. We are making the rich richer and the poor poorer. Over the last ten years, the average income from investments has nearly doubled in rate. This means that the increase in the gross national product goes largely to benefit the wealthy, and to give them more and more of a stranglehold on the economy of the country and of the world—rather than going to relieve the suffering of the poor.

Twenty percent of the Negroes in the United States have an average annual income of only $250 a year. This very small income leads to illness and suffering.

The governments of the world show their immorality by wasting 10 percent of the world's production on militarism. Doctors should be opposed to war. . . .

You know, there was a time when the religious leaders were the scientists and physicians, and in some societies they still are—the shaman, the witch doctor. Then a sort of specialization came in.[17]

In after years, with lay readers and audiences especially, Pauling might not mention the Canadian psychiatrists but would attribute the origin of his concept of orthomolecular medicine itself to biochemist (and vitamin C enthusiast) Irwin Stone, who in a 1966 letter to Pauling had recommended that he take ascorbic acid so as to live longer and feel

better (Chapter 11). Both origins of Pauling's interest in vitamin therapy, which are highly relevant to his subsequent work, occurred around the same time and complemented each other.

Pauling in 1992 gave a talk to the Schizophrenia Foundation, entitled "Vitamin C: The Key to Health." In it he recaptured the thinking process that led him into vitamin research.

I didn't have any intention of getting heavily involved in the vitamin field, except that I became interested in my own health and what vitamins might do for me. But then it happened that I came across the work of Abram Hoffer and Humphry Osmond. I had been working for ten years on the molecular basis of mental disease when I read what Hoffer and Osmond had done in Saskatchewan, giving vitamin B_3 to schizophrenic patients—and also ascorbic acid (vitamin C). At first it didn't make a very great impression on me. I just didn't know much about vitamins.

After perhaps a week, I thought, "There's something odd here. Here I am sixty-six years old and I've never heard of substances like these vitamins having the properties these vitamins had." The point was we know that a little pinch—a few milligrams of vitamin C a day—is enough to keep people from dying of scurvy. A little pinch of nicotinic acid (niacin) is enough to keep people from dying of pellagra. But Hoffer and Osmond were giving a thousand times or ten thousand times these amounts, and they had powerful physiological effects other than keeping people from dying.

Well, I knew a moderate amount about drugs, and I knew that a drug that has some effectiveness against a disease is more effective the larger the amount that you give. So I knew that doctors, being very sensible people, and the medical profession, having studied these problems for a long time, felt that if a person is seriously ill and there is a possibility of saving his or her life by use of the drug, then you should give as much of the drug as you can in order to increase the probability of saving the person's life. And the upper limit is that if you give too much of a drug, you hurt or kill the person. So you have to compromise by giving somewhat less than the toxic or lethal dose, and this is what physicians do.

What about the vitamins? Here are these extraordinary substances that have very powerful physiological effects when they are given in small amounts to people, enough to keep people alive. And yet there is a range of a thousandfold or ten-thousandfold in intakes over which these substances can be taken without killing the person. And I thought, "There are two questions that we might ask about a vitamin." First, how much does a person need in order not to die of the corresponding deficiency disease? We know the answer to that. That's the RDA, the recommended dietary allowance set by the appropriate government or semigovernmental bodies. The RDAs are enough to keep people barely alive in ordinary poor health.

The other question is, over this great possible range of intakes, where is the intake that puts people in the best of health? As a scientist, I thought, "Well, I'd better look in the literature and find out what the answers to this question are for different vitamins." And I thought, "It's astonishing that I haven't read about the optimum intake of vitamins in the sixty-three years that I have been reading." When I looked in the literature, I couldn't find anything. So I thought, "It's odder still that this important question hasn't been answered." The result, of course, is that for the last twenty-five years I've spent a large amount of time and energy trying to answer it.[18]

Pauling realized that vitamins—micronutrients essential to maintaining health in all living creatures—must be obtained from food by animals (plants can synthesize them), with one noteworthy exception. Vitamin C is formed endogenously within most mammals, reptiles, fish, and birds. Usually this synthesis takes place in the liver, but sometimes in the kidneys. Only humans and the other primates, the guinea pig, a bat species, and several bird species do not make their own supplies of ascorbic acid as needed. This *special need* in humans to get vitamin C from the external world therefore qualified it, in Pauling's view, as a *molecular disease.* The broad implications of this evolutionary quirk intrigued him.

Although Pauling later said that he first publicly proposed the concept of orthomolecular medicine in 1968, in fact he mentioned it at least twice in 1967. Pauling revealed this new approach to medicine in a popular lecture he gave at Montreal's Expo 67. After presenting a brief history of structural chemistry and then charting his own pathway through biomedical research, he introduced the concept of orthomolecular medicine. Then—not unexpected for him at that or any other time —he launched a critique of the prevailing economic system and its effect on public health.

I am convinced that the general plan of attempting to control disease and improve health by varying the concentrations in the body of substances that are normally present and are required for life is a good one.

I believe that we can reduce the amount of human suffering caused by disease very greatly through discoveries that have been made and that will be made in the future. I am disturbed, however, by the fact that we are not using our present knowledge effectively. Statistical studies show that there is a correlation between health, as expressed in life expectancy, and income; in general the life expectancy is increased as the income of a population becomes greater, there being an increase in life expectancy by about three years for each doubling of the average income of a population. This general correlation leads us to expect that the people in the United States would be the healthiest in the world, but they are not. In several European countries, where the average income is between one-quarter and one-half that of the United States, the life expectancy is from three to six years greater, rather than less. I surmise that this deviation from the correlation is to be accounted for as resulting from the maldistribution of medical services in the United States. This maldistribution itself may be related to the great spread of income. Of the 200 million people in the United States, 20 percent, 40 million, live on only 5 percent of the national income. They constitute the poor people. Another 5 percent of the national income is distributed to or got by only 650,000 people, the rich people. The ratio of incomes of the rich and the poor is sixty to one. The state of health of the 40 million poor people is a reflection of their small income and expresses itself in low life expectancy.[19]

Pauling also explicitly used the term "orthomolecular" in the title of a communication—"Orthomolecular Somatic and Psychiatric Medicine" —that was to be read at a major conference, the Thirteenth International Convention on Vital Substances, Nutrition, and the Diseases of Civilization. His message was published in January of the following year. The meeting was being held in Luxembourg in September 1967, but Pauling, who had planned on going, found himself moving his household to San Diego, where he had accepted a temporary professorship in chemistry. In his paper Pauling discussed what he meant when originating this new field of medical research and practice.

I have reached the conclusion . . . that a general method of treatment of disease, which may be called orthomolecular therapy, may be found to be of great value and may turn out to be the best method of treatment for many patients.

Orthomolecular therapy is the treatment of disease by the provision of the optimal molecular constitution of the body, especially the optimal concentration of substances that are normally present in the human body and are required for life. The adjective "orthomolecular" is used to express the idea of the right molecules in the right concentration. The word may be criticized as a Greek-Latin hybrid, but I have not thought of a better word.

I believe that orthomolecular therapy may have a special value in the treatment of mental disease. The functioning of the mind is dependent on its molecular environment, the molecular structure of the brain. . . . The proper functioning of the mind is known to require the presence in the brain of many different substances.[20]

In retrospect, the evolution of this Pauling-designated field of orthomolecular psychiatry into nutritional medicine generally seems inevitable. Pauling himself did not really separate mind and body, so it was suitable for him to apply orthomolecular medicine to both *psyche* and *soma,* as the article's title had it.

Pauling's most prominent early paper on the subject, "Orthomolecular Psychiatry," was published in *Science* in April 1968 with a provocative subtitle: "Varying the Concentrations of Substances Normally Present in the Human Body May Control Mental Disease." Here are a few of his conclusions and the summary relating to his biochemical analysis of the possible genetic causes of schizophrenia.

I T has been suggested by Huxley, Mayr, Osmond, and Hoffer on the incidence of schizophrenia in relatives of schizophrenics, that schizophrenia is caused by a dominant gene with incomplete penetrance. They suggested that the penetrance, about 25 percent, may in some cases be determined by other genes and in some cases by the environment. I suggest that the other genes may in most cases be those that regulate the metabolism of vital substances, such as ascorbic acid, nicotinic acid or nicotinamide, pyridoxine, cyanocobalamin, and other substances. . . . The reported success in treating schizophrenia and other mental illnesses by use of massive doses of some of these vitamins may be the result of successful treatment of a localized cerebral deficiency disease involving the vital substances, leading to a decreased penetrance of the gene for schizophrenia. There is a possibility that the so-called gene for schizophrenia is itself a gene affecting the metabolism of one or another of these vital substances, or even of several vital substances, causing a multiple cerebral deficiency.

I suggest that the orthomolecular treatment of mental disease, to be successful, should involve the thorough study of and attention to the individual, such as is customary in psychotherapy but less customary in conventional chemotherapy. In the course of time it should be possible to develop a method of diagnosis (measurement of concentrations of vital substances) that could be used as the basis for determining the optimum molecular concentrations of vital substances for the individual patient and for indicating the appropriate therapeutic measures to be taken. My coworkers and I are carrying on some experimental studies suggested by the foregoing considerations, and hope to be able before long to communicate some of our results.

The functioning of the brain is affected by the molecular concentrations of many substances that are normally present in the brain. The optimum concentrations of these substances for a person may differ greatly from the concentrations provided by his normal diet and genetic machinery. Biochemical and genetic arguments support the idea that orthomolecular therapy, the provision for the individual person of the optimum concentrations of important normal constituents of the brain, may be the preferred treatment for many mentally ill patients. Mental symptoms of avitaminosis sometimes are observed long before any physical symptoms appear. It is likely that the brain is more sensitive to changes in concentration of vital substances than are other organs and tissues. Moreover, there is the possibility that for some persons the cerebrospinal concentration of a vital substance may be grossly low at the same time that the concentration in the blood and lymph is essentially normal. A physiological abnormality such as decreased permeability of the blood-brain barrier for the vital substance or increased rate of metabolism of the substance in the brain may lead to a cerebral deficiency and to a mental disease. Diseases of this sort may be called localized cerebral deficiency diseases. It is suggested that the genes responsible for abnormalities (deficiencies) in the concentration of vital substances in the brain may be responsible for increased penetrance of the postulated gene for schizophrenia, and that the so-called gene for schizophrenia may itself be a gene that leads to a localized cerebral deficiency in one or more vital substances.[21]

Note that the vital substances Pauling proposed to optimize are vitamins —and that Pauling did not deny the correlative value of having some talk therapy as part of the orthomolecular treatment, thereby leaving some space for the psychiatrist, who would function as both physician and counselor. But in an article in *Nutrition Today,* Pauling forcefully expressed his belief that since most mental diseases were probably attributable to biochemical causes, the prevalent psychoanalytic approach to treatment was misguided.

THE idea that insanity is of metabolic origin is not new. Indeed, as far back as 1881, a distinguished lecturer on the subject at the School of Medicine at Edinburgh University, Dr. J. Batty Tuke, defined insanity as "a disease of the brain inducing disordered mental symptoms." This is a definition that is gradually finding renewed acceptance today. . . .

In the earliest practice of medicine, the corporeal character of insanity was generally admitted. It was not until the superstitious ignorance of the Middle Ages had obfuscated the scientific, though by no means always accurate, deductions of the earlier writers that any theory ascribing a purely psychical character to insanity was proposed.

In the 1920s an unfortunate episode occurred—the introduction into psychiatry of psychoanalysis. Most of our present psychiatrists have been trained in this school, which traffics in a jargon of special meanings and meaningless words. We know now that psychoanalysis has failed, and scientists are turning back to the concepts of the corporeal origin of psychiatric disorders that were current ninety years ago. No doubt this change in thinking is due in part to the advent of new drugs that are useful in altering cerebral metabolism. The fact that such drugs have an effect on the psyche is evidence that these disorders are not purely psychic but have a metabolic base. More people are coming to realize that the brain, the organ through which mental phenomena are made manifest, is vulnerable to outside substances. The existence of an insane mind in a metabolically healthy brain is unlikely. The brain is the most sensitive of all our organs to variations in its molecular composition, and mental symptoms of avitaminosis often are observed long before any physical symptoms appear. I do not say that supportive psychotherapy is without value. When mental disorders, such as schizophrenia, are looked upon as metabolic dysperception, and the hope is instilled that rectification of the molecular imbalance will lead to improvement, the patient usually can be helped.[22]

Ever the crusader, Pauling began challenging and confronting the medical establishment—at first mainly the psychiatrists, who used nonphysiological methods of treatment (such as intensive psychoanalysis), or extreme but sometimes only temporarily fashionable physical forms (such as lobotomy, electroshock, and insulin-coma therapy), or the potent synthetic drugs likely to produce drastic side effects, even toxic reactions. He couldn't understand why these physicians were unwilling to try out orthomolecular psychiatry's approach, on the chance that it would at least benefit some patients. "I believe," he asserted, "that a psychiatrist who refuses to try the methods of orthomolecular psychiatry, in addition to the usual therapy, in the treatment of his patients is failing in his duty as a physician."

Pauling postulated that acute forms of mental illness, such as psychotic breaks or the periodic decompensating episodes in paranoid schizophrenia, were caused by depletion of particular brain chemicals— occurring perhaps under stressful circumstances. If minor symptoms had been discounted or neglected, an orthomolecular solution might have corrected the brain-fluid condition before a serious episode ensued. These were things that he already knew about molecular diseases generally. There should be ways to diagnose such conditions early.

Psychiatrists, as well as psychologists, other psychotherapists, and many nutritionists, considered Pauling an interloper in their own domains. For the most part they dismissed Pauling's introduction of orthomolecular medicine—if they had heard at all about his assertions regarding megavitamin therapy to alleviate mental disorders. The time was not quite right yet for regarding mental health as a potential field for serious research into biochemicals (such as the neurotransmitters and neurohormones serotonin, norepinephrine, and the endorphins) that affect thinking and feeling states. As the twenty-first century approaches, though, Pauling's fascination with the biochemistry of the brain seems prescient indeed.

By the late 1960s Pauling was moving determinedly into his next research arena: human nutrition in general—and vitamin C in particular.

Eleven

Vitamin Crusader

Having the right molecules in the right amounts in the right place in the human body at the right time is a necessary condition for good health.

By 1970 Linus Pauling was ready to make strong statements to the public, not just to the medical profession, about the promise of nutritional medicine for the prevention and treatment of disease. Several years earlier he had indicated that orthomolecular psychiatry, which advocated establishing the brain's proper biochemical equilibrium by megavitamin administration, might be extended into orthomolecular medicine generally. The keystone of his theory, apart from having the right balance and forms of macronutrient foods—proteins, carbohydrates, fat—was having the correct, preferably optimum intake of other nutritional factors, including micronutrients. These were the vitamins, minerals, amino acids, and other vital substances utilized in enzyme formation, metabolism, and the additional thousands of energy-generating, cell-building, and other complex physiological processes that sustained life.

All the essential nutriments had to come from outside sources, from foods and, if needed, special dietary supplementation. Some of these elements and compounds were obviously more crucial than others, and Pauling envisaged new equipment and techniques for testing bodily fluids to determine orthomolecular deficiencies, imbalances, and dysfunctions so as to intervene in disease progression or obtain what he called optimum health.

Pauling, always the earnest educator, by now had gathered up abundant intriguing health information that he wanted to impart widely to the public as well as to health-care practitioners. Already experienced and well known as an effective communicator on nuclear and international-peace issues, he was in a better position than most nutritional researchers to claim people's attention. The possession of two unshared Nobel Prizes boosted his credibility.

Dr. Pauling as benevolent prophet emerged whenever he talked about advances in chemistry, science, and peace. The seer's dark side appeared, however, whenever he insistently dealt with the bad news. Nuclear weaponry and fallout, warfare, injustice, the neglect of public health and other societal problems would sometimes elicit jeremiads from him, on platform and in print. Increasingly he sought the American people's eyes and ears when television crews aimed their cameras at him. He always saw the value of pulling public opinion over to his side; people would then apply pressure on officialdom to correct some intolerable situation or change a prevailing attitude.

A terrific enthusiast whenever he had something positive to present about his findings or opinions, in 1970 Pauling brought the gospel of vitamin C. And for the rest of his days he would preach it to everyone, private citizens and government officials—executives, legislators, and bureaucrats—journalists and physicians, whether they wanted to hear or not. He transferred some of the evangelistic passion he felt for promoting chemistry and peace to proselytizing on behalf of orthomolecular medicine—with vitamin C as the core tenet. At a time when pharmacological researchers were expanding the testing and use of powerful synthetic compounds to stabilize or reverse different forms of physical or mental illness, Pauling looked to biological substances, found to occur naturally in the human body, for solutions—a viewpoint he maintained for the rest of his life. His earlier, intense interest in creating artificial enzymes for the treatment of molecular diseases departed.

How had Pauling's new involvement in orthomolecular medicine progressed? In the autumn of 1967 he had become a research professor in chemistry at the University of California's young campus in San Diego (UCSD). There he undertook various lecturing assignments and some graduate-student mentoring. He began working closely again with a former biochemistry student of his from Caltech. Arthur B. Robinson, who as a graduate student had earlier participated in the hydrate-microcrystal-anesthesiology research, was now on the faculty at UCSD. The partnership seemed a good match. Both men were fascinated with the diagnostic potential in computer-assisted equipment, increasingly accessible to researchers. And they had considerable experience with isolating, identifying, and measuring different substances in fluids and tissues. By combining special software with sophisticated lab equipment, they aspired to identify and analyze hundreds of separate biochemical markers within human body fluids (blood, urine, saliva), even volatile compounds in breath or urine vapor.

New laboratory methods could serve as diagnostic tools and would also provide explicit measurements whereby suitable biochemical adjustments could be determined for deficiencies in vital substances, particularly the vitamins and minerals, or for excesses in some, such as iron, copper, and vitamins A and D, that might adversely affect health.

Once Pauling moved over to expounding the principles of orthomolecular *somatic* medicine, he, Robinson, and their colleagues began writing scientific papers and grant proposals to explain what they were already doing or wanted to do—funding permitting. Pauling's interest in psychiatric disorders continued, but now he saw mental functions as connected with the chemistry of body physiology.

I T has been suggested that the treatment of disease should be carried out by adjustment of the concentrations of substances in the human body so as to duplicate those concentrations which correspond to optimum health and that this adjustment should be carried out whenever possible by the administration of compounds normally found in the human body. The administration of the right molecules in the right amounts is called "orthomolecular therapy." It has further been suggested that a quantitative assessment of the state of molecular health of the human body, "orthomolecular diagnosis," might be made in a practical way by the quantitative analysis of the normal constituents of human urine followed by a simple type of computerized pattern recognition. Human urine has been found to contain several hundred normally volatile compounds that are easily separated by gas-liquid chromatography. . . . We have, therefore, devised an apparatus that is suitable for routine quantitative analysis of about two hundred volatile constituents of human urine.[1]

In 1969, two years after he went to UCSD, Pauling transferred his research activities and grants (he had secured several from NIH and its mental-health offspring, NIMH) from San Diego up to Stanford University, in the San Francisco Bay area. Expressing dissatisfaction with the conservative policies and politics of the state-run university system (ex–Hollywood actor Ronald Reagan was currently governor of California), Pauling said he preferred to return to a private institution of higher learning and research. He persuaded Art Robinson to take a leave of absence from his tenured position at UCSD and join him.

Geography probably influenced Pauling's decision as well. Information-exchange opportunities and financial backing related to the molecular-biology research and biotechnology development that most interested him now would be available in the Santa Clara Valley, increasingly known as Silicon Valley. Since the area was fast becoming the innovative biotech and computer center of the world, both creative mental energy and venture-capital investment levels were high. If government didn't want to support Pauling's diagnostic projects significantly, new industries might. There was a potential for patents here.

The last major segment of Linus Pauling's research and public-education work, concentrating mostly on orthomolecular medicine, had begun. Taking place on the San Francisco peninsula, it would last for twenty-five years. The Paulings bought a house in Portola Valley, close to Stanford, and settled in for the duration.

In this close connection with orthomolecular medicine, Pauling was also becoming knowledgeable about human nutrition in general and vitamin C in particular. During the latter years of his psychiatric studies Pauling had noted, starting with others' clinical observations, that the blood and urine of many mental patients showed subnormal levels of vitamin C and various B vitamins, particularly niacin, pyridoxine, and cobalamin (vitamins B_3, B_6, and B_{12}), indicating either poor absorption, metabolic errors, or an increased uptake or demand for them. For whatever physiological reasons, doubtless these deficiencies were associated with brain malfunctions that affected perception, thinking, emotions, and behavior. Moreover, it was also shown that symptoms of mental disorder and emotional distress were relieved in some patients who regularly received megadoses of particular vitamins, through as yet unknown biochemical or electrochemical corrections taking place in the brain.

Pauling was gathering laboratory and clinical studies, past and current, about different vitamins and minerals. Apt anyway to regard most psychological or psychiatric disorders as essentially biochemical and physiological in origin, he concluded that both mental and physical health problems might be successfully treated through administering particular combinations of micronutrients and also regulating diet, with regimens designed for individual need. Physician Thomas Addis was actually doing this with his kidney patients two decades earlier—Pauling's personal introduction to taking vitamin supplements and also to the still-embryonic precept of orthomolecular medicine that encouraged some food substances while eliminating or reducing others (Chapter 6).

But people's requirements in certain health conditions—as, perhaps, for much larger doses of particular vitamins than the normal diet could supply—would need to be determined by some methodology. Pauling was intrigued with nutritional biochemist Roger J. Williams's proposed theory of biochemical individuality. (Williams had discovered both pantothenic acid, vitamin B_5, and folic acid; his brother Robert was the codiscoverer of thiamine, vitamin B_1.)

Relocated at Stanford, Pauling and Robinson, with other researchers, continued developing tests and laboratory equipment for identifying key biochemical markers for particular health conditions. Pauling expected, of course, that some or even many of them would ultimately indicate early manifestations of genetic disorders and degenerative diseases that might be corrected, or at least kept under control, through orthomolecular adjustment. "Wrong" molecules could be as crucial as the "right" ones.

Pauling described this rationale to mental health professionals in a 1970 talk, "Orthomolecular Psychiatry." Dealing with more than mental problems, it expressed Pauling's bafflement over physicians' dismissal of nutritional medicine and amply indicated the special interest he was now taking in vitamin C.

I feel that the use of substances normally present in the human body for improving health of human beings, and especially their mental health, has been unjustifiably ignored by the medical profession for some thirty or thirty-five years now and that the possibilities of improvement in the health of the American people and of other people in the world by improved nutrition are truly great. It is astounding to me that the medical profession has paid so little attention to these possibilities during the last few decades.

It is difficult for me to understand why this has come about. There was enthusiasm about vitamins and about nutrition for a rather short period of time, beginning about 1910, when vitamins were first clearly recognized and when it was generally accepted that diseases such as scurvy and beriberi are not the result of the presence of a toxic substance of some sort in certain foods, which could be neutralized by other foods, but are rather the result of the absence in certain foods of vital substances, the vitamins.

The essential amino acids also were discovered to be vital substances of this sort, required for life and health.

The enthusiasm about vitamins may have been overly great for a while and the failure of vitamin therapy in some cases may have caused a disenchantment that really was not justified. . . .

I would like to know how to find out what the optimum rate of intake of these vital substances is.

Professor Williams has reported studies made with guinea pigs showing that for optimum growth the amounts of vitamin C required by different guinea pigs varied by as much as twentyfold. He has said that surely human beings are more heterogeneous and the range of requirements of human beings for vitamin C is greater than over a factor of twenty. I feel that we can say the same thing about vitamin B_3, a very important substance which has many functions in the human body, and also vitamin B_6, which is known to serve as a coenzyme in many enzyme systems. I feel sure that the needs for vitamin B_3 in different human beings are different.

I have decided, on the basis of the evidence presented by Irwin Stone, that there is very strong evidence now that most human beings are suffering from hypoascorbemia, a mild sort of deficiency of ascorbic acid in the blood—perhaps it is wrong for me to call it a *mild* sort. The point that I call to your attention is that I believe that for all or almost all human beings the amount of vitamin C that is contained in the food is less than the optimum amount and that the state of health of almost all human beings is not so good as it would be if they were to ingest a larger amount. . . .

What are we going to do about this problem in the future—the problem of finding out what the needs are of individual human beings for these important foods?

At the present time, I feel that it is essential that empirical methods be used. . . .

We know that these substances have physiological activity, that many of them are involved in the functioning of the brain, that they cause biochemical reactions to take place in the human body such as to change the molecular environment of the mind; and we know by observation, from the reports that we have heard today and from others that have been published, that often these changes in the molecular environment of the mind are such as to lead to improved behavior. . . .

Professor A. B. Robinson, my associate, and our coworkers have been working on the problem of finding out what the molecular structure of a human being is, finding out how he handles vitamins, what happens to the vitamins that he ingests—does he utilize them in the same way as other human beings. . . .

We are now trying to develop an instrument such that it does not take four hours (as it does at present) to determine quantitatively the amounts of two to three hundred substances in a sample but will take only a few seconds. . . .

It is possible now to make a new attack on the whole problem of nutrition in relation to the health of human beings and especially of mental health. . . .

It is soon going to be possible to answer this question and many similar questions. It is required only that scientists and physicians, medical investigators, have an open mind about such matters as the value of vitamins, that they not be inhibited by old and false ideas that have been handed on by the past generation of physicians and nutritionists to the present generation.

Fortunately, the younger generation of physicians and of students generally is less gullible than those of earlier times—more open minded. I think that the attitude of the young physicians and the students of today gives us hope for the future.[2]

By 1970, then, Pauling was fully embarked on the last great crusade of his life. Though still distracted by the need to keep speaking out against the war in Vietnam, he was poised to launch his promotion of vitamin C, to elevate its importance above that of all other vitamins. His enthusiasm came not only from reports he read but also from his own direct experience of its benefits. He believed that this essential micronutrient should be ingested in generous amounts by all humans. His descriptions of its uses, which expanded as the years went by, made it appear to be almost a panacea. When Pauling talked or wrote about ascorbic acid, the scientists' name for vitamin C, his words and tone had almost a missionary zeal.

Even in his nineties, Pauling was still telling interviewers and public audiences the story of his conversion to taking vitamin C megadoses himself. He told it many times in different ways; usually he credited Irwin Stone with the personal revelation, whereas Hoffer and Osmond's psychiatric work with vitamins pertained more to his originating the concept of orthomolecular medicine. "My Love Affair with Vitamin C," published in *Health Care USA* in 1992, catches some of the long-sustained evangelism. The article's title, though doubtless created by an editor with a flair for attention-getting headlines, has a certain authenticity—though Pauling himself avoided using sentimental language, except sometimes when he talked about his wife. (Ava Helen would never have permitted him to have an affair, even in his head, with anybody or anything . . . except, perhaps, for vitamin C. After her death in 1981 he became devotedly celibate, reserving his passions mostly for calculating the bonding arrangements in molecular structures or for investigating and promoting ascorbic acid.)

I had begun taking an ordinary vitamin-mineral supplement, containing the RDAs (recommended dietary allowances) in 1941. It was not until 1966, however, that I began to develop an interest in vitamin C and other vitamins taken in far larger amounts than the RDAs.

During a talk in New York City, I mentioned how much pleasure I took in reading about the discoveries made by scientists in their various investigations of the nature of the world, and stated that I hoped I could live another twenty-five years in order to continue to have this pleasure. On my return to California I received a letter from a biochemist, Irwin Stone, who had been at the talk. He wrote that he was sending me copies of some papers he had just published, with the general title "Hypoascorbemia, a Genetic Disease," and that if I followed his recommendation of taking 3,000 milligrams of vitamin C (ascorbic acid, sodium ascorbate, potassium ascorbate, calcium ascorbate), I would live not only twenty-five years longer, but probably more.

The 3,000 milligrams per day that he recommended is fifty times the RDA. Soon after, I began to feel livelier and healthier. In particular, the severe colds I had suffered from several times a year all of my life no longer occurred. After a few years, I increased my intake of vitamin C to ten times, then twenty times, and then three hundred times the RDA (now 18,000 mg per day).

Two arguments Stone presented to support his thesis that the proper physiological intake of vitamin C is fifty or more times the RDA especially impressed me. Most vitamins are required exogenously by all species of animals. Presumably an early animal, eating plants as food, was getting enough of these vitamins in the food to come close to satisfying its needs, and the process of evolution led to the loss of the ability to manufacture them in cells of its body. Vitamin C is an exception, however.

Almost all animal species—dogs, cats, cows, elephants, and so on—have continued to synthesize ascorbate. A likely reason for this is that animals require more ascorbate for good health than plants do. For one thing, ascorbate is required for hydroxylation reactions, and is used up during these reactions. A most important reaction of this sort for animals is the conversion of procollagen to collagen. Collagen is the principal structural protein of animals, which strengthens the blood vessels, the skin, the bones, the teeth, and other tissues; whereas plants use a carbohydrate, cellulose, for this purpose, and hence do not need a large amount of ascorbate in order to synthesize collagen. One might conclude that for an animal to depend upon plants as a source of vitamin C would result in a deficiency of this vitamin.

The second fact that impressed me is that animals manufacture large amounts of ascorbate. The amount manufactured is approximately proportional to body weight, and converted to the weight of a human being, ranges from about 2,000 to 20,000 milligrams per day. Irwin Stone concluded that human beings with an average diet are accordingly all suffering from hypoascorbemia, a deficiency of ascorbate in the blood and tissues.

The amount provided by an average diet for a human being is about 60 mg per day, the RDA.[3]

Pauling established a good working relationship with Irwin Stone, who was in the process of writing his classic book, *The Healing Factor: Vitamin C Against Disease* (1972). Stone's earlier promise of twenty-five additional years was better than fulfilled; Pauling lived three years beyond it. (Sometimes Pauling got carried away, saying that Stone had offered him fifty years—especially toward the end of his life, when the twenty-five years had just about been used up.)

Again and again, in years to come, Pauling would discuss the neat logic in Irwin Stone's and other researchers' conclusions, as he did here in "Good Nutrition for the Good Life."

PLANTS manufacture vitamin A, vitamin B_1 (thiamine), vitamin B_2, B_6, and other vitamins for themselves. Animals require these substances exogenously, and we can ask why. I think the answer is this: In the early days of the existence of animals, they had inherited from their plant ancestors the machinery for making these important substances. But they were eating plants, and the plants manufactured these substances, so they were getting a supply of them in their food. It may well be that the amount of vitamin A that animals were getting was just about as much as they needed—close to the optimum. Now if a mutant came along that had suffered a genetic deletion, losing the genes that are involved in producing the enzymes that catalyze the reactions leading to the synthesis of vitamin A, the mutant would still have vitamin A from his food, but he would be a streamlined animal, not burdened by the machinery for *making* vitamin A, and in the competition with a more slowly moving competitor who was handicapped by this machinery, he would win out. The situation would be the same with thiamine, riboflavin, pyridoxine, and other vitamins. I believe that this is what happened, and that this is why all animals require the vitamins.

But this didn't happen with vitamin C. . . .

What happened? I think with little doubt that these are not separate mutations for human beings and gorillas and rhesus monkeys and other primates, but rather a single mutational loss—a common ancestor 25 million years ago, living in a tropical valley where the fruit foods were especially rich in vitamin C (providing 10 or 15 grams per day for a body weight of 70 kilograms), underwent a mutation. The mutant lost the machinery for making the vitamin C and was correspondingly stream-lined and able to compete and as a result the mutant won out, and we are all descended from this mutant, who suffered this unfortunate accident. As long as our ancestors stayed in this area they were getting enough vitamin C. When they moved into temperate and subarctic regions, the food available contained less vitamin C, and they began to suffer from scurvy.

One measure of good health is resistance to disease. There have been over a dozen carefully controlled studies carried out on a comparison of vitamin C tablets and placebo tablets in blind trials, with respect to the incidence and severity of the common cold. Every one of these studies carried out with people exposed to cold viruses by casual contact with other people has shown that vitamin C has protective value. There is no doubt about it. In fact, if, in addition to taking regular doses of vitamin C, you carry a supply with you and increase the intake at the first sign of a cold, or even other illness, taking 10 to 20 grams during the first day, and then tapering off, you can stop the cold. Many cold medicines make you feel better, but they don't prevent the cold from developing. Vitamin C will do this. Not only that, but vitamin C prevents other diseases.[4]

Occasionally Pauling also cited a third source besides Hoffer-Osmond and Stone for his professional introduction to the wonders of ascorbic acid.

ALTHOUGH my wife became interested in vitamin C long ago, shortly after it was discovered, I first became interested in it in 1966, when I read papers by VanderKamp, who showed that the metabolism of this vitamin by chronic schizophrenics is unusual. My wife and I carried out some tests on ourselves and other people, to see what fraction of a large dose of vitamin C that is taken by mouth is excreted in the urine. Dr. Robinson, who was then assistant professor of biology at the University of California, San Diego, and I then began studying patients with acute schizophrenia, in comparison with other people. We verified that there is something unusual in that many schizophrenics have a very low body content of some other vitamins. My interest in vitamin C then spread to its effect on other diseases. I found, when I read the medical literature, that a number of investigators had reported that a high intake of vitamin C gave some protection against the common cold. I also found that the medical authorities and the authorities in the field of nutrition denied that vitamin C had any such value in protecting against the common cold or any other disease, except scurvy. This discovery caused me to write my book *Vitamin C and the Common Cold*, which was published in 1970.[5]

Actually, though, the precipitating factor that persuaded Pauling to write the bestselling book may have been an individual's reaction to his talk "Medicine in a Rational Society," given in 1969 to medical students and faculty at the opening ceremony of the new Mt. Sinai Medical School in New York City. Already in lectures Pauling was extolling the virtues of vitamins in general and vitamin C in particular. Now, as was customary in the Vietnam War period, Pauling first urged the young physician-candidates to join the revolt against the current immoral militarism of their government and to move toward establishing a rational society, in which they would contribute to progress in health care and biomedical research. Then, after discussing molecular diseases, he asserted that "we should be paying more attention to the natural vital substances, the vitamins and essential amino acids." He went on to describe the condition of avitaminosis—diseases caused by deficiency of particular vitamins in many mental patients. Finally, he praised the vitamin with which his name would henceforth be associated.

A large intake of vitamin C, 1 gram a day or more, has been reported to be of value also in accelerating wound healing and recovery from infection, including the common cold. These reports have been rejected by most medical authorities, who have contended that the usually recommended daily intake, about 50 milligrams, is enough for every person. It is my opinion that we do not know what the optimal daily amount of vitamin C is. I think that for most human beings it may lie between 1 gram and 5 grams per day, far more than the usually recommended 50 milligrams per day; and, moreover, as Professor Roger J. Williams has emphasized, that there may be large differences in the needs of different human beings.

We should know what the optimal daily amounts of the various vitamins are. This is a medical problem that should be attacked and solved.[6]

What happened next was the beginning of a private feud made public, which lasted the rest of Pauling's lifetime, for a quarter century. (It still goes on posthumously, when Pauling is no longer able to talk back to his vocal, perennial adversary—and to others like him who objected to Pauling's strong beliefs and work in numerous areas concerning science and society.) Pauling had laughed about the start of this altercation when talking in the late seventies during the "Plowboy Interview" for *Mother Earth News* magazine.

THEY had invited several people to speak at this ceremony, so I only had about ten minutes—maybe fifteen—to speak, but in my short talk I mentioned the value of vitamin C in preventing colds as something important to medicine, to health in general.

Well, one of the professors who attended the ceremony wrote me a very strongly worded letter attacking me for having made the statement about vitamin C. He said, "Do you want to support the vitamin quacks that are bleeding the American public of hundreds of millions of dollars a year?" and he asked: "Can you show me a single double-blind study that indicates that vitamin C has any more value than a placebo in fighting colds?"

I wrote to this fellow and told him that no, I couldn't show him any studies, but that I hadn't really looked at the literature, either. And I didn't pursue this for two or three months . . . but it kept bothering me. Finally—after several months—I got around to checking the medical literature . . . and I found *six* double-blind studies, every one of which showed that vitamin C did in fact have more value in preventing colds than a placebo. And by "double-blind" I mean that neither the people dispensing the pills nor the people receiving them knew which pills contained the vitamin C—and which ones were the placebos—until the end of the study. The records were kept, in other words, by a third party.

So again, I wrote to this fellow—I didn't expect to go beyond this, you see—and said that I had found that there were several studies backing me up, one of which was a 1961 study—written in German—by Ritzel. I gave him the reference to Ritzel's paper, and I thought that would end the matter.

The professor wrote back and said that he was too busy to hunt up the reference to Ritzel. Well, I made a Xerox copy of Ritzel's paper and mailed it to him, so he wouldn't have that excuse [laughter]. Then he wrote to me, saying, "I am not impressed by the work of Ritzel." I wrote back and said, "I'm not impressed by your saying that *you're* not impressed by the work of Ritzel. After all, the boys in Ritzel's study who got the vitamin C had only a third as much illness due to colds as the boys who got a harmless placebo . . . and the numbers in the paper have high statistical significance. You can't just say you're 'not impressed' by the work . . . you have to have a reason."

Well, the professor wrote and said that Ritzel didn't give the age of his subjects, nor their sex . . . which happens to be untrue. I wrote the professor and said that because I had lived for a year and a half in Germany after receiving my Ph.D., I could read Ritzel's German without trouble, and it seemed clear to me that Ritzel said that his subjects were all boys in their teens. So then the professor wrote to me and said, "Well, there are two ski camps in the study, and perhaps Ritzel gave the vitamin C to the boys in one ski camp and the placebo to the boys in the other ski camp, and maybe the camps were different in some way, and . . ." Well, I wrote to Ritzel about that, and he said—essentially— "How silly can you get?"

Here's this man, this professor—I didn't identify him when I wrote my book—Victor Herbert, who to this day keeps writing papers and giving speeches saying that no one benefits from taking extra vitamins . . . and he won't even look at the evidence.

The upshot of this whole thing is that I finally became sufficiently irritated by this fellow that I decided I ought to do something about it. So I sat down one summer—here, downstairs in my study—and in two months wrote a book *Vitamin C and the Common Cold*.[7]

Pauling's new book, which he began to write in the spring of 1970, would be simple, readable, and informative, with short chapters. His regular textbook publisher, W. H. Freeman in San Francisco, agreed to issue it. Pauling had already written two popular books before—*No More War!* and *The Architecture of Molecules*.

Most physicians and medical researchers had paid no attention yet to the probability that vitamin C supplementation would improve human health. Another Nobelist, Albert Szent-Györgyi (who won the Nobel Prize in Physiology or Medicine for discovering ascorbic acid), also believed that doctors were misleading the public about vitamin C's value. He maintained that the vitamin was fundamental to life and took at least one gram per day. So in *Vitamin C and the Common Cold* Pauling used a tactic with already proven value. Just as he had taken the nuclear-testing issue to the American public, he now set out to deliver his message about vitamin C for general consumption. What sickness was most annoying in its frequent recurrence to Americans? The virus-engendered common cold, of course. Pauling could not understand why physicians did not widely recommend the taking of vitamin C to their patients as an excellent prophylactic.

The book was published in December of 1970. Much to everyone's surprise, it swiftly became a bestseller. Readers picked up Pauling's own enthusiasm for vitamin C. Pauling was in demand as a speaker and a

TV-program guest. The press interviewed him and quoted him. This type of popularity he had never experienced before, and he appeared to enjoy himself hugely. Inevitably, in this new mania there was a run on vitamin C supplies in drugstores that continued until fine-chemical and pharmacological manufacturers—in the United States, Europe, and Japan—expanded their facilities for mass production of the synthetic form of vitamin C, which could be sold cheaply. (The process usually involves chemical conversion from sugar, which as a compound contains the same elements—carbon, oxygen, and hydrogen—but in different proportions. Vitamin C is now the leading seller as a single vitamin.)

Pauling was pleased that once again he had stirred up a storm over an issue that was not being sufficiently examined and discussed, either by the supposed experts or the public. He now became vitamin C's major champion worldwide; his book had other English editions, published in the United Kingdom and India, and was translated into eight foreign languages, including Japanese. (Pauling always enjoyed great popularity in Japan, in part because of the antinuclear stance that brought him and Ava Helen on peace pilgrimages to Hiroshima.)

Pauling parlayed this newly won popularity into promoting the new field of orthomolecular medicine that he had originated—to encourage the public to look more closely at nutritional factors in their lives and also to expect their physicians to pay better attention to vitamins' possible importance, sometimes a crucial one, in health and recovery. Long accustomed to talking to ordinary Americans through his writings and speeches, Pauling now fitted well into the proselytizer's role. He was not preaching as yet to the assembled faithful; instead, through focused education he hoped to convert the ignorant and convince the disbelievers. He wanted his message to resonate in people's lives. Seemingly indefatigable, he wrote and talked. For almost a quarter century he told—to whoever asked, and sometimes to those who did not—about his own personal experiences with vitamin C, about various research studies done and how they should be interpreted, and about vitamin C itself: what it is, why people can't make it in their own bodies (the absence of this mechanism qualified the condition as a molecular disease), its long human history with the deficiency disease called scurvy, how much might be taken for what, and when. But take it, he urged: it is unlikely to harm you in large amounts, may cut down on the frequency, severity, and duration of viral colds and flus, and should make you feel healthier, more energetic, more cheerful. He himself, at the age of seventy plus, seemed the very picture of prime mental and physical vigor—a walking,

talking advertisement for the wonders of ascorbic acid, particularly perhaps to the middle-aged and elderly.

In later talks and writings, Pauling would also discuss alternatives to
plain ascorbic acid—sodium ascorbate and calcium ascorbate—which
could be taken by people whose stomachs were bothered by its acidity.
He also advocated the use of high-dose intravenous vitamin C in certain
health conditions, including emergencies—which was effectively done
by various physicians whom he knew or corresponded with. He warned
against the rebound effect when high regular doses of ascorbate are
abruptly discontinued, with the hypothetical potential for causing
scurvy symptoms. And he discussed the remote possibility of megadoses
causing oxalate kidney stones, exacerbating iron-storage disease, and any
other adverse side effects other than a mild laxative action.

In espousing what he called preventive nutrition, Pauling encouraged
other micronutrients' gaining favor with the public. Because he had
become the best-known spokesman for nutritional medicine generally,
many people—in the United States and elsewhere in the world—came
to believe he was actually either a physician or a nutritionist, a highly
informed authority about all facets of nutrition. Many of them were
totally unaware of his distinguished scientific work and of his antinuclear and peace advocacies. (As with most of the other research fields he
had entered in his long career, he never took a class in nutrition, let
alone obtained an advanced degree that would enable him to claim bona
fide professional status. Thus, properly credentialed nutritionists and
physicians might dismiss him as a charlatan.)

When the Vietnam War began winding down in 1973, Pauling could
give even more time and energy to his public campaign of promoting
vitamin C. As with his past crusades, in advocating vitamins Pauling
always managed to sound utterly convinced of the grand purpose of this
self-assigned mission on earth. He was rather like a populist politician
perennially campaigning for some high public office, eager to gain adher-
ents who would vote for his cause against the establishment's experts,
who apparently would never agree with his stand—unless the American
people forced them to. Now the entrenched authorities Pauling fought
were physicians and medical associations, who generally ignored, ridi-
culed, or actively opposed the orthomolecular and basically megavitamin viewpoint that Pauling promoted. Government health agencies also
declined to heed Pauling's insistence that funding be given to basic
vitamin research that would determine the optimal intakes of the vitamins and other nutrients, not just the minimal amount needed to avoid
a deficiency disease. NIH, however, did help considerably for a while to

fund Pauling's ongoing inventions of orthomolecular diagnostic research methods and tools.

By 1973 Pauling had the novel experience of being sought by mainstream publishers. The *Saturday Evening Post* invited him to write an article about nutrition. And *True* published an article, "Man and His Health: The Strange Case of Vitamin C." Before reviewing for readers the FDA's attempts—even then—to impose a limit on the dosage amount of any vitamin put into a single tablet or capsule, he revealed his irritation over how the FDA's commissioner had dismissed his book. (Health-care authorities would find, over the years, that Pauling was not an easy adversary. He could marshal legions of protesters to bombard legislators with letters, telegrams, and phone calls damning the FDA's hyperregulatory position—far more readily than in his years as a radical peacenik.)

In 1975 Dr. Pauling visited the Senate again, but this time not in response to a subpoena summoning him for a harsh interrogation. Instead, he appeared before the Senate Subcommittee on Health, chaired by Edward Kennedy, to testify officially on behalf of vitamins and effectively argue the case for dietary supplements, which the FDA intended to treat as prescriptive drugs when pills contained more than the RDA. He built his calm, rational argument upon his knowledge of and experience with vitamin C, as the best example for continuing an open market on all such helpful nutritive substances. At the same time, he succinctly summarized the salient virtues of vitamin C, citing a number of specific research studies demonstrating that far more than the currently established RDA was needed to avoid or ameliorate the common cold, assist the healing of wounds and burns, control back trouble, decrease the incidence of heart and vascular diseases, inactivate viruses and bacteria, and achieve mental acuity and the feeling of overall well-being. He also pointed out the ironic fact that animal nutritionists had established that animals, whether or not they synthesize ascorbic acid, require much more vitamin C to sustain health than is considered necessary for humans.

Then Pauling concluded:

I believe that the vitamins are important foods, and that the optimum daily intakes of vitamin C and other vitamins, leading to the best of health, are much larger than the present recommended dietary allowances. I believe that the American people should not be hampered in their efforts to improve their health by an intake of vitamins approaching the optimum intake. The proposed FDA regulations would operate in a serious way to make it difficult for the American people to obtain these vitamins, by classifying them as drugs in daily amounts greater than the U.S. RDAs. I accordingly support legislation that will prevent the Food and Drug Administration from carrying out this unwise action.

The values of the RDA for various vitamins have been set by the Food and Nutrition Board by consideration only of the amounts needed to prevent death or serious illness from a dietary deficiency. No serious consideration whatever has been given to the question of the optimum daily intake, the amount that leads to the best of health. . . .

For several years I have taken 6,000 milligrams of vitamin C each day. I take it as pure crystalline L-ascorbic acid or as 1,000-milligram tablets. If the FDA regulations were to go into effect, I would be put to added trouble and expense. I might be restricted to buying tablets containing the U.S. RDA of 100 milligrams, so that I would have to swallow sixty of these tablets each day. This would mean ingesting a large amount of filler and binder in the tablets, the filler and binder constituting a larger fraction of the 100-mg tablets than of the larger tablets. Also, the small tablets are more expensive, per gram of vitamin C, than the larger tablets. An alternative would be for me to go to the trouble of getting a physician to prescribe large doses of vitamin C for me. Aside from the trouble of getting the physician to do this, I would have to pay his fee, and would also have to pay the customary higher price for prescription items. I am sure that the new regulations, if they were to go into effect, would operate to the detriment of the health of the American people.

As a scientific investigator, I am interested in carrying on research on medical problems, including the problem of determining as reliably as possible the values of the optimum intakes of vitamins and other nutrients. Classification of vitamin C as a drug would in my opinion work a serious hardship on the research effort at a time when it should be most encouraged. The regulations about research on the effects of drugs on human subjects, which in my opinion are quite proper, would operate in an unnecessarily restrictive way to hamper research in the field of nutrition, especially research on the improvement in general health of people accompanying an increased intake of various vitamins.

As a consumer, I am concerned about the misrepresentations and overpricing that have existed in connection with the sale of vitamin C and other vitamins. Even at the present time, some vitamin C preparations are being offered for sale at prices as much as one hundred times those of essentially equivalent preparations. Advertising is often misleading in suggesting a difference in vitamin C depending upon whether it contains wild rose hips or is a preparation of pure crystalline L-ascorbic acid. Preparations presently available under the name Rose-hip Vitamin C may contain less than one percent of rose-hip powder, with less than one-hundredth of one percent of the vitamin C coming from rose hips. The proposed FDA regulations do not in my opinion establish an effective mechanism for protecting the consumer against the abuses of misrepresentation and overpricing. A more direct approach, I believe, lies in implementation and enforcement of strict requirements about truth in advertising and initiation of a broad-based campaign of consumer education. It should be required, for example, that the actual amounts of the various components of each vitamin C preparation be stated on the label. Requirements of this sort about truth in advertising and labeling would be extremely helpful in eliminating some of the most serious consumer abuses.

I believe that the expression "Recommended Dietary Allowance" used by the Food and Nutrition Board and by the FDA is misleading, in that the RDAs are not the amounts that should be recommended as providing the best of health, but are only the amounts, probably much smaller, that prevent death or serious vitamin deficiency disease.[8]

Although congressional legislators subsequently trimmed various attempts by the FDA to strictly regulate the over-the-counter vitamin products along with other dietary supplements and herbal products, the issue is bound to keep recurring. And the RDAs—or whatever new terms and abbreviations are contrived for the numbers set by the Food and Nutrition Board of the National Academy of Sciences as standard minimal daily intakes of various micronutrients—will continue both to gauge and to reflect how the medical profession regards vitamin therapy: with doubts and suspicions, whether justifiable or not. Among Pauling's last published letters to the editors were statements defending vitamin supplementation. He felt that this was a liberty almost as fundamental to the American people as those in the Bill of Rights.

In 1973 Pauling was poised to make yet another move—the fourth within a decade. His professorship at Stanford University had begun in the fall of 1969. When transferring from University of California, San Diego, he had brought some research grants and research personnel (notably Art Robinson) with him. He soon discovered, though, that he would not be given nearly enough laboratory space to satisfy his needs. Stanford administrators weren't able or willing to provide more than a makeshift arrangement. (Doubtless some of them, too, were alarmed at Pauling's concurrent, highly visible, and vocal anti–Vietnam War activities; at the time of his arrival nationwide protests were reaching a fever pitch. There was always the danger that wealthy donors would end their generosity if offended by what they considered undignified, radical-leftist behavior. Despite the old peace warrior's fame and brilliance, there were definite limits to conservative alumni's tolerance, as Pauling's Caltech experience showed.)

Additionally, Pauling's taking up so publicly and ardently with vitamin C soon after arriving at Stanford struck many biomedical science professionals—physicians in particular—as material evidence that he was getting senile and silly. Stanford's science-research administrators declined to support Pauling's keen desire to investigate ascorbate as a vital adjunct to good health. When his popular book on the common cold was published, he was well past the standard retirement age. To many of them Pauling probably seemed well over the hill in terms of producing serious and meaningful original research that would reflect creditably on their institution. With his fixation on vitamins he was simply becoming, according to this jaundiced view, almost a quack or snake-oil salesman. Where was the brilliant structural chemist of yesteryear? they wondered—failing to comprehend how Pauling's hyperactive humanism could lead him into ventures with their own intrinsic scientific and rational validities, as it had with nuclear testing. In this case, improving health, public and private, was his paramount concern.

Pauling's requests for sufficient laboratory room for his ambitious orthomolecular research—involving a building extension or even a temporary structure—were not met. As a double Nobelist and world-renowned scientist he expected better treatment. He and his colleague Robinson began planning an exit from academia. In 1973 Pauling was in his early seventies. Scarcely planning a peaceful retirement in some new, resort-type senior citizens' colony, he continued to ponder many research ideas and was eager to implement them. Above all he intended to continue investigating vitamins and other orthomolecular substances. In a closely correlated way, he and Robinson also wanted to perfect their methods of testing bodily fluids for diagnostic information useful in treatment or prevention.

The solution to Pauling's dilemma at Stanford was becoming obvious: he, Robinson, and other committed researchers would create their own biomedical research facility, which someday might even acquire the academic accreditation needed in offering advanced degrees. They felt assured about securing continuous NIH and other government grants for health research, as well as financial assistance from local biotech companies that might benefit from their discoveries and inventions.

Robinson secured a building in Menlo Park where they could begin setting up their independent research endeavor, while Pauling would continue in his well-salaried Stanford professorship until the following year, 1974, when he would officially retire. Then, with articles of incorporation in hand, in May of 1973, Pauling, Robinson, Keene Dimick (a manufacturer of laboratory equipment), and several others constituting the board of trustees (including Linus Pauling, Jr., M.D., Pauling's eldest son, a psychiatrist who practiced in Hawaii) applied for official standing as a nonprofit corporation with the state of California and for tax-exempt status with the Internal Revenue Service. They succeeded in both quests.

The name chosen for the new organization, the Institute of Orthomolecular Medicine, would be changed in the following year to the Linus Pauling Institute of Science and Medicine (LPI). The original name did not have the attraction of the principal scientist's identity; besides, the term "orthomolecular medicine" still baffled people. (Passersby, seeing the sign, sometimes dropped in to ask what it meant.) Although Pauling persistently used the term—just as he always favored the abbreviation "CIT" to saying or writing "Caltech"—"orthomolecular medicine" never quite caught on with health professionals or the public. It is usually called and explained now as nutritional therapy or nutritional medicine, even by some LPI-associated people aware of a certain flaky implication. Unfortunately, neither alternative name precisely conveys the biochemical and molecular aspects that Pauling intended to stress when originating the concept.

Pauling clearly intended now to give much of his time to pondering problems in human health and poring over reports and data from other researchers' studies, while continuing to design provocative experiments that his own colleagues could perform under his guidance. He had learned much from the work of theoretical physicists. During the past century they (Einstein was the prime example) had rarely gone into laboratories, but by working often in solitude in ways that were both cooperative and competitive with other physicists and mathematicians, they had developed new ideas that radically reshaped scientific thought, thereby to change the world. Pauling believed that the same attention to deep and speculative thinking should take place in medicine.

THE time has come for those responsible for the steady progress of medical education and health care to give thoughtful consideration to the establishment of a professorship of theoretical medicine in each of our medical schools. . . .

Simply to sit and think about a tremendously complex and difficult medical problem such as cancer is no longer necessarily a waste of time. It might have been in earlier decades, when we were considerably less well informed about the structures and functions of the human body than we are today. Today, in our present, more advanced state of knowledge, it is important for us that some gifted minds should be allowed to ponder undisturbed, to attempt solutions of unsolved problems by the exercise of reason alone.

If they were, we would be led to a new appreciation of known facts, and the concept of orthomolecular medicine would be given a chance. We might discover that disease can be controlled to a significant extent by the ingestion of appropriate amounts of nutrient substances. A proper intake of vitamin C might prove to be the single most important factor in human health.[9]

At LPI Pauling's research increasingly shifted away from orthomolecular psychiatry into investigating the aging process, particularly after the NIMH funding ran out. It seemed reasonable to apply orthomolecular theory and tactics to delaying the inevitable slowing down or disruption of crucial biochemical transactions that brought on decreased enzymatic action, cellular oxidation, DNA deterioration, malabsorption of nutrients, and inefficient metabolism—resulting in the age-associated acceleration of degenerative diseases like atherosclerosis and cancer.

When Pauling in 1960 had published his lecture "Observations on Aging and Death," he early established an interest in these two interconnected phenomena in humans, biochemically and statistically. In 1974 he produced a long paper, "The Process of Aging." The publication is intriguing for various reasons, such as an early mention of the free-radical theory of aging and disease (to be countered by antioxidants such as vitamins C and E), protein cross-linking, and other biochemical and molecular explanations that are now common in both research and popular literature. Pauling was eager to expand orthomolecular research into aging so as to indicate certain life-extension possibilities in preventive nutrition. Frequently he gave talks to civic groups in both the United States and abroad on issues pertaining to the health of the elderly, especially after How to Live Longer and Feel Better was published in 1986.

In time, Pauling hoped, the diagnostic techniques that he and his colleagues at LPI were working on would provide assays of key markers for such conditions as micronutrient deficiencies, abnormal proteins in blood and urine, and excesses of substances such as iron, lead, and other metals—which, along with hazardous environmental chemicals, could have dire physiological consequences that would speed up the aging process. The cumulative effects of exposure to radiation, Pauling's fixation in the fifties and early sixties, was also to be factored into the equation. Additionally (as he had done in a Caltech-era publication a decade and a half earlier), Pauling issued stern warnings about the health consequences of smoking. "If you want to lead a miserable life," he would say, "take up smoking." Smokers at least could take a lot of vitamin C to help counter the destruction of supplies of ascorbate in blood and tissue.

In 1976 a revised and expanded *Vitamin C, the Common Cold, and the Flu* was published in time for the public to prepare for the swine-flu epidemic predicted for the coming winter. The new preface commented on vitamin C's potential in combating viral infections, whether a deadly influenza or some much milder forms nicknamed flu.

Pauling never claimed that vitamin C would prevent or cure everyone's colds; it would help only a percentage of people in a significant way part of the time. He also knew from his own and others' experiences that taking megadoses of vitamin C early—1 gram (1,000 mg) per hour at the first signs of an impending cold—often defeats the virus.

After the publication of the initial *Vitamin C and the Common Cold,* many people who got colds and didn't like them—especially those habitual sufferers—began trying Pauling's recommendations for themselves. A good number of them contacted Pauling after discovering that vitamin C definitely reduced the number, duration, and severity of the colds that customarily afflicted them. People with lifelong histories of high susceptibility to colds and flus said that, most amazing, they rarely got them at all now. They might also report additional benefits apparently effected by vitamin C. It appeared to boost immune function, promote wound healing and the repair of broken bones, renew energy, relieve allergies, reduce pain, neutralize various toxins, and even help to overcome substance addictions. It seemed to quickly clear up intractable infections, bad bruises, gingivitis, poison ivy, infectious mononucleosis —and more serious sudden or chronic afflictions. These and other curative phenomena had, after all, been presented by Irwin Stone in *The Healing Factor,* based on laboratory research, clinical findings, and personal experience.

"Those are anecdotal case histories; they don't mean anything!" researchers and physicians usually said when Pauling spoke of personal communications that came in continually from the public. He knew the absence of clinical data or double-blind trials made them unacceptable as valid research. Still, he always liked to hear from and talk to people themselves who had stories to tell, because their individual experiences confirmed his own position regarding either the multifaceted value of ascorbate or the value of other micronutrients to people with individual biochemical health needs. He believed that compiling personal reports would provide cumulative evidence of the therapeutic merits—or, conversely, certain drawbacks—that should be investigated by researchers at LPI or elsewhere.

From the mid-1940s to the early 1970s, Pauling as a scientist had devoted about half of his time to the cause of promoting world peace. In the last twenty or so years of his life, he took much time and energy away from the pure theoretical research that he really preferred doing, to perform a perpetual social service for bettering human health, whether universally for humankind or for individuals. Pauling's dedication to orthomolecular medicine derived from his firm belief in the scientist's duty to contribute to society—if possible, in some significant and lasting way. He is now probably better known and remembered by the public for this effort than for all others.

Twelve

The Nucleus of Controversy

Every person on earth is essentially in poor health.
It's called "ordinary good health," but it ought to be
called "ordinary poor health."

Through the years people would often phone or write to Linus Pauling, perhaps in despair over a diagnosis recently made of their own or a loved one's health indicating that the condition was terminal—beyond the ability of modern medicine to cure. They might have then recalled or been told by someone else about Pauling's advocacy of vitamin C's amazing properties, particularly in treating cancer.

Some persons even came by the institute's headquarters in hopes of obtaining an audience with the great Dr. Pauling. Occasionally someone actually got invited in to see him, for he always kept open the possibility of personal contacts with strangers. He was especially available to those with a challenging health problem to discuss, for which he might then suggest an unorthodox approach based on biochemical and nutritional principles. He'd tell them to stay in touch with him and report on any progress. A man of compassion, he depended largely on the gatekeeping skills of his secretary-assistant, Dorothy Bruce Munro, who from 1973 on served as his public-relations representative—while also typing his manuscripts, transcribing dictated audiotapes, handling his voluminous correspondence, making appointments, arranging complicated travel itineraries, and setting up speaking dates and interview schedules.

One of Dr. Pauling's dearest and longest-held dreams, according to Mrs. Munro—as Pauling always called her, formally, just as he was always "Dr. Pauling" to her—was to have a permanent clinic where people could be helped by physicians and other health professionals dispensing orthomolecular medicine. At such a clinic all sorts of useful information could be gained from investigative studies and controlled trials. Tests would also be run on blood and urine samples to try out and then perfect the high-tech devices and analytical methods on which LPI researchers were diligently working. Such specialized clinical measurements and experiments, using megavitamin therapy done with willing patients, weren't normally attainable, of course, in conventional hospitals and clinics. (Probably too there was a nostalgic aspect to Pauling's desire, harking back to his life-saving visits during the 1940s to Thomas Addis's informal renal clinic at Stanford Hospital in San Francisco. There, as befitted Addis's political liberalism, patients of all socioeconomic levels mingled democratically; they were also encouraged to watch the kidney-affecting procedures conducted on the resident population of experimental rats, which related to their own conditions.)

In 1974, as part of LPI trustees' rapid-expansion plan for when Pauling would leave Stanford, it was decided to open an outpatient clinic. Dr. J. Frank Catchpool accepted the position of clinic director. His credentials were excellent. As a young physician he had spent six years in West Africa working in the famous clinic at Lambaréné run by Dr. Albert Schweitzer. There Catchpool initially met Pauling upon his arrival in 1959 with Ava Helen on a two-week visit to discuss peace and anti-nuclear strategies with the revered philanthropic physician, who was likeminded about these issues. Later, Catchpool had joined Pauling at Caltech to take part in biomedical research. British born and trained, in the past year he had interned at the San Francisco General Hospital to qualify for the license requisite in American medical practice.

When the Linus Pauling Institute's clinic opened its doors in late 1974 at its quarters in Menlo Park, prospective patients who learned about this new development began arriving in droves at the doorstep. Many had terminal cancer or were schizophrenics; they had heard that Dr. Pauling and his institute specialized in treating these diseases. Some traveled from afar, as if coming to the American Lourdes, expectant of medical miracles to be wrought by Dr. Pauling or his fellow redeemers. Often these desperate people had no funds for food, lodging, or transportation, let alone the ability to pay for examinations, treatment, and vitamin supplies. If not out of their minds upon arriving, they tended quickly to become so.

Pauling, Catchpool, and their associates felt sympathetic and charitable toward the crowds jamming their small facility, but they soon became desperate themselves. In setting up this idealized plan for combining research and treatment, nobody seems to have anticipated the cost or commotion that running such a clinic would entail. An undercapitalized LPI could not afford to continue on with the bedlam and financial drain, and adequate medical-liability-insurance coverage was prohibitive. Not surprising, in about six months the clinic ingloriously folded. Yet ever afterward among some members of the public and even a few health professionals, the notion of a still-operating LPI clinic persisted.

So Pauling realistically abandoned the hope of doing human clinical studies directly under the institute's auspices. From then on, he depended upon clinical and epidemiological research done elsewhere by others. If it involved suggestions or more active involvement from him, his name might appear as coauthor on published findings.

In the ensuing years, the Linus Pauling Institute lurched from one financial crisis to another. Its unconventional, vitamin-oriented, independent-minded approach to biomedical research tended to rule out the possibility of government grants beyond the early ones, relating to diagnostic techniques, that had been transferred from Stanford. Considerable support began to come in from private foundations and from individual donations, largely through direct-mail solicitations and the cultivation of wealthy donors. In 1979 the board of trustees, at Pauling's request, terminated Robinson's employment. Pauling was aware of mounting staff disgruntlement with Robinson, such as over an LPI association with an anticancer-research study that used wheat grass and other plant products—considered likely to amplify LPI's reputation for flakiness among serious scientists. (Robinson's ensuing lawsuit was not settled for several years, during which time LPI moved from its original Menlo Park offices to a former factory site in Palo Alto.) Emile Zuckerkandl, who had worked with Pauling earlier, succeeded Robinson as director.

At LPI Pauling and his fellow scientists, removed from direct clinical work with human subjects, largely conducted hands-on biomedical investigations with in vitro body fluid and tissue studies and in vivo research using animals, primarily mice and guinea pigs—like humans, the latter (conveniently) have the same anomalous defect and do not synthesize ascorbic acid. They also used computers and other electronic instruments to assist in measurements and conclusions for their laboratory experiments. The comparatively small research staff managed to accomplish a great deal under somewhat limited circumstances. By 1995, more than five hundred scientific publications altogether had come from their work—which included not only several hundred in-house laboratory findings but also computational studies of DNA sequencing and collaborative associations with researchers elsewhere. Published too were Pauling's ongoing crystallographic and nuclear-structure studies and delvings into aspects of metallurgy, such as devising superconductive substances with possible applications for sensitive diagnostic equipment —work in which he was often joined by LPI researcher Zelek Herman. The institute also periodically hosted scientific conferences.

The most productive and cooperative connection that Pauling ever formed for doing clinical studies was with a surgeon in Scotland, Dr. Ewan Cameron. This close association, begun when they were mutually addressing the prospect of treating cancer with ascorbic acid, led to a succession of intriguing clinical results—as well as rejections, triumphs, and frustrations. (There is a still-abiding promise in this vitamin C therapy.) Their partnership, spanning two decades, also provided Pauling with discouraging insights into the workings of the research-funding system in the United States and of an entrenched medical establishment that for the most part refused to consider his postulations regarding cancer prevention and therapy. This predicament, along with his other proposed orthomolecular uses of ascorbic acid or other micronutrients, Pauling would discuss in a 1985 article entitled "Problems Introducing a New Field of Medicine."

ONE of the problems in introducing the field of orthomolecular medicine to the medical profession and to the public is that there is a great amount of misrepresentation about the toxicity of vitamins. This emphasis on toxicity and possible harmful side effects from intakes of the vitamins higher than the recommended dietary amounts is found both in articles in the popular press and in some scientific and medical journals, and it probably represents a bias based upon a lack of knowledge, especially about recent investigations. An example is provided by an episode occurring a few years ago when a small boy swallowed all of the contents of a bottle of vitamin A and began to suffer nausea and headache. He was taken to the hospital, treated, and was then released. The physicians prepared an account of the episode for a medical journal, and news stories about this poisoning by vitamin A were published in the *New York Times* and hundreds of other newspapers. Every day someone dies of aspirin poisoning, but no newspaper stories about the deaths are printed. When a small boy becomes sick enough from eating a large number of vitamin A tablets to require treatment in a hospital, however, the physicians themselves and the newspapers consider the fact worthy of publication.

A great amount of work still remains to be done on the general problem of getting information basic to orthomolecular medicine. What are the optimum concentrations of orthomolecular substances in the body fluids? What are the optimum intakes of the various vitamins? . . .

Because of their remarkably low toxicity, vitamins should be considered as a class of substances different from drugs, which in general have high toxicity and are often prescribed in doses close to the lethal level. The water-soluble vitamins are especially innocuous. . . .

The possibilities of improving the health of the aging population through an increased intake of vitamins may be especially great, but it is also likely that younger people can benefit by improved nutrition through the use of vitamin supplements. It may turn out that the greatest contribution to public health made during the past decade or two has been the recognition of the value of vitamins ingested at the optimum intake levels.[1]

Pauling sometimes pointed out that, ironically yet helpfully, vitamin C megadoses actually seem often to counteract the terrible nausea and other side effects suffered by patients when undergoing chemotherapy, which involves the intermittent intravenous administration of a toxic substance. In some patients, he said, ascorbic acid may even "potentiate" the drug's curative actions. However, most oncologists seem unaware of this synergistic effect and are dubious when patients report good results.

In a 1977 report in *Executive Health* Pauling briefly presented an overall perspective on cancer in public health.

CANCER is the cause of nearly one-fifth of all deaths in the United States. Each year about five hundred thousand people in this country develop cancer, and most of them die of the disease. It is especially important that control of cancer be achieved, because the amount of suffering associated with cancer is much greater than that for most other diseases. It is for this reason that the federal government has emphasized research on cancer, and has allocated several hundred million dollars per year for cancer research, reaching $800 million this year.

Despite the great amount of money and effort expended in the study of cancer, progress during the last twenty years has been slow. A significant increase in survival time after diagnosis was achieved about twenty-five years ago, largely through improvements in the techniques of surgery and anesthesia. During the last twenty years some improvement in treatment of certain kinds of cancer has been achieved, mainly through the use of high-energy radiation and chemotherapy, but for most kinds of cancer there has been essentially no decrease in either incidence or length of time of survival after diagnosis, and it has become evident that some new ideas are needed, if greater control over this scourge is to be achieved.

Pauling then introduced the topic of using vitamin C as a prophylactic and treatment, giving some historical background.

ONE new idea is that large doses of vitamin C (L-ascorbic acid, sodium ascorbate, calcium ascorbate) may be used both to prevent cancer and to treat it.

The biochemist Irwin Stone in his 1972 book *The Healing Factor: Vitamin C Against Disease* discussed the early reports that doses of vitamin C of 1 to 4 grams per day, sometimes given together with an increased intake of vitamin A, seemed to have value in controlling cancer in some patients. This work was done largely by German physicians in the period between 1940 and 1956. Despite the indication that these doses of vitamin C were of value in the treatment of cancer, the early studies did not lead to a thorough examination of the possible use of vitamin C in treating cancer. Some favorable results were also reported in studies with animals, but the early work in this field too was not followed up to any extent in the cancer research program.

In 1951 it was reported that patients with cancer have usually a very small concentration of vitamin C in the blood plasma and in the leukocytes of the blood, often only about half the value for other people. This observation has been verified many times during the last twenty-five years. The level of ascorbic acid in the leukocytes of cancer patients is usually so low that the leukocytes are not able to carry out their important function of phagocytosis, that is, of engulfing and digesting bacteria and other foreign cells, including malignant cells, in the body. A reasonable explanation of the low level of vitamin C in the blood of cancer patients is that their bodies are using up the vitamin C in an effort to control the disease. The fact that the level is low in cancer patients suggests that they should be given a large amount of the vitamin in order to keep their bodily defenses as effective as possible. Despite this reasonable argument, however, little attention was paid to vitamin C for prevention and treatment of cancer until 1971.[2]

In 1971 Pauling himself had begun to speculate in public about this use of vitamins—as in this portion of a speech he gave at the dedication of the new Ben May Cancer Laboratory at the University of Chicago's Pritzker Medical School. Having said some things about different studies with vitamin C and the cold virus and schizophrenia, he moved on to cancer.

THERE are in the medical literature a number of papers in which the relation of nutrition to cancer is discussed. These papers have been largely ignored.

It is known that vitamin C is required for the synthesis of collagen by the body. It is required for wound healing. It is required for preserving the strength of blood vessels. Vitamin C is an antioxidant, and vitamin E is also an antioxidant. Tissues in the human body can be damaged by oxidation of some of the molecules that constitute them, especially the unsaturated hydrocarbon side chains in cell membranes and the constituents of the interstitial ground substance. Ascorbic acid is essential for the synthesis of the collagen fibrils in the ground substance, and it may well function in other ways to strengthen the ground substance and to prevent the infiltration of tissues by cancerous growths. Ascorbic acid is known to have antiviral activity, and some cancers involve viruses. Preserving the integrity of the tissues by proper nutrition could prevent cancer cells from penetrating through the tissues, and contribute to the prevention of the development of cancer and the spread of cancer. So far as I am aware, this approach to the cancer problem has been almost entirely neglected during recent years. Only a few physicians, especially Dr. W. J. McCormick of Canada and Dr. F. R. Klenner of North Carolina, have made trials of ascorbic acid for the prevention and control of cancer. They have reported that it has value, but their results have been discounted or ignored.

I believe that there are great possibilities for the future. If proper nutrition were to decrease the number of cases by 10 percent, this would be a most important contribution, saving fifteen or twenty thousand lives in the United States per year. I believe that proper nutrition can control cancer to a much greater extent than 10 percent. Nutrition, vitamins used in their proper amounts—these are matters that the scientists and medical men have neglected too long. I hope now that the role of nutritional factors in the attack on cancer will be thoroughly investigated.[3]

Pauling had based a portion of his argument for trying out vitamin C in the prevention and treatment of cancer on information he had obtained from a book published five years earlier by a Scottish physician.

I read this book and was impressed. In it, Dr. Cameron expressed the idea that malignant tumors liberate the enzyme hyaluronidase, which attacks the hyaluronic acid in the intercellular cement that holds the cells of the body together. It weakens them so that the malignant tumor can then infiltrate the normal tissues. He said that if he could find a way of increasing the production by the body of a natural physiological hyaluronidase inhibitor, that might control the malignant tumor by inhibiting the action of this enzyme. He tried for years by giving mixtures of hormones to terminal cancer patients to achieve this goal. . . .

[In the Chicago lecture] I made of his argument that it is also the fibrils of collagen which strengthen the intercellular cement. We know that vitamin C is required for the synthesis of collagen, so it might well be that the control of cancer might be achieved by giving ascorbic acid in the proper amounts to the patient.[4]

What happened next was that Pauling soon received a letter from the book's author, Ewan Cameron, chief surgeon of Vale of Leven Hospital, Loch Lomondside, Scotland.

DR. Cameron read an account of my address, and wrote asking how much vitamin C should be given to patients with terminal cancer. My recommendation was that he give 10 grams per day. This is 220 times the usually recommended intake of the vitamin. Fortunately, vitamin C is a remarkably innocuous substance; much larger amounts, 100 to 200 grams per day, can be taken by intravenous infusion without serious side effects.

Dr. Cameron has recently stated that his immediate reaction to the idea that vitamin C could have value against cancer was sheer incredulity. He has said that he was a conservative Scottish surgeon, and that the people around him in Vale of Leven Hospital were conservative Scottish medical people. It seemed quite ludicrous to them to suggest that this simple, cheap, harmless powder, which could be bought in any drugstore, could possibly have any value against such a bafflingly complex and resistant disease as cancer. The solid logic of the arguments persisted, however, and Cameron decided that the cancer problem was such a serious one that it seemed worth making a trial of the idea, with patients with advanced cancer. If the vitamin C did the patient no good, at least it would do him no harm.[5]

Pauling met his perfect companion in professional risk taking in this Scottish physician who proved willing to administer vitamin C to hospital patients.

DR. Cameron decided to try that, which shows how brave he is. Because not many physicians or surgeons are willing to risk their reputations in this way. In the fall of 1971 he cautiously administered ascorbate to the first terminal cancer patient, and then to the others. And very soon, I judge, he became enthusiastic. He has now [1979] had more than 750 such patients to whom ascorbate has been administered as the only therapy, and others to whom it has been given as an adjunct to other methods of treatment of cancer.[6]

By 1977 Pauling was able to talk about the promising results of Cameron's clinical work, many of whose patients lived longer—some even still survived—and reported an improvement in well-being. Cameron placed emphasis on the positive subjective responses of his patients. (Cameron had been made a nonresident fellow of the Linus Pauling Institute, so that his research was also, by extension, LPI's.)

WHEREAS he had been disappointed in his trials of various hormones, he immediately felt that the treatment with vitamin C was of considerable benefit to the patients, and during the next five years he gave the vitamin in large doses to more than one hundred patients with advanced cancer, almost all of them being patients for whom the conventional methods of treatment had been tried and found to be of no further benefit. He and his coworkers published several papers on their observations. In one paper they reported that the vitamin C seemed to control pain quite effectively, so that patients who had been receiving large doses of morphine or diamorphine could stop taking the narcotic drug. He also published a detailed report on the first fifty patients with advanced cancer to be treated with large daily doses of vitamin C, and a paper on one patient who seemed to recover completely from cancer when treated with vitamin C, in whom, however, the cancer returned when the intake of vitamin C was stopped, and who again recovered completely when the treatment with vitamin C was resumed. This patient continues to take vitamin C, 12.5 grams per day, and after three years seems to be in excellent health.

In the recently published report a comparison is made of [the] one hundred patients . . . and one thousand other patients, ten matched controls for each of the ascorbate-treated patients (same kind of cancer, same age, same sex). The one thousand controls were given the same treatment as the ascorbate-treated patients except for not receiving ascorbate. They were patients in the same hospital and were treated by the same physicians and surgeons.

The result of this study was that the ascorbate-treated patients have lived on the average over four times as long as the matched controls.

Sixteen of the one hundred ascorbate-treated patients have lived more than one year, whereas only three of the one thousand controls have lived that long. Moreover, although all of the one thousand controls have died, sixteen of the one hundred ascorbate-treated patients are still alive, and a dozen of them seem to be free of disease.

The vitamin C treatment seems to have some favorable effect for all kinds of cancer. . . . The largest effect was observed with cancer of the colon. The thirteen ascorbate-treated patients with cancer of the colon lived more than seven times as long, on the average, as their 130 matched controls. The next largest effect, a nearly sixfold increase in survival time, was observed for patients with breast cancer, followed by a five-times increase in survival time for patients with cancer of the kidney, and smaller ratios for those with cancer of the bladder, the rectum, the bronchus, the stomach, and the ovary (only a twofold ratio for cancer of the ovary). For several other kinds of cancer the average survival time for the nineteen ascorbate-treated patients was three times that for the 190 matched controls. There is accordingly some indication that the ascorbate treatment is more effective for cancer of the colon, the breast, and the kidney than of the others, but it is not certain that this difference exists. At the present time the conclusion can be drawn that a high intake of vitamin C is beneficial for all patients with advanced cancer.

Dr. Cameron has now given the ascorbate treatment to more than two hundred patients with advanced cancer and to a smaller number with cancer in earlier stages. He is planning to publish reports on the later groups of patients as soon as possible.

There is [also] considerable evidence from epidemiological studies that an increased intake of ascorbic acid decreases the incidence of cancer. . . .

From these various observations it seems likely that vitamin C is effective both in preventing cancer and in treating it.

There is evidence that a high intake of vitamin C operates to increase the effectiveness of essentially all of the protective mechanisms in the human body. Its action in strengthening the intercellular cement . . . [may enable it] to inhibit the action of the enzyme produced by cancer cells to attack the intercellular cement. It is also known that vitamin C is involved in the various immune mechanisms in such a way that an increased intake makes them more effective in attacking and destroying the malignant cells. The antiviral and antibacterial effectiveness of vitamin C may also be of significance with respect to cancer. A detailed understanding of the various mechanisms of action of vitamin C against cancer will not be obtained until much additional research has been carried out. . . .

At the present time there are some physicians who prescribe vitamin C in the amounts used by Ewan Cameron in Vale of Leven Hospital or in smaller or larger daily amounts for patients with cancer, in addition to the conventional treatment or some other treatment. In addition, there are many physicians and surgeons who are willing to administer vitamin C, both intravenously and orally, to those patients who request that it be used. Carefully controlled trials of the value of vitamin C for cancer of different kinds and at different stages in the progress of the disease will without doubt be carried out during the next few years. Until these trials have been carried out, the work of Ewan Cameron remains the best source of information about the value of this vitamin in relation to cancer.[7]

During the 1970s Pauling visited Cameron several times in Scotland to see the cancer patients for himself and to review Cameron's data. They prepared a first report on this work, which Pauling then sent as a matter of course to the *Proceedings of the National Academy of Sciences,* which did not subject articles submitted by its noted scientist-members to the peer-review process customary in other prestigious journals. So Pauling was shocked when the *PNAS* editorial board refused to publish it, on the grounds that the study hadn't been properly conducted in the conventionally accepted manner, with double-blind controls. Pauling told them that the clinical circumstance was not designed for or conducive to that purpose; he maintained that the recorded results should be of sufficient interest to researchers and physicians. In spite of Pauling's appeal for reconsideration and mustering up protests from other academy members, this unprecedented editorial decision held.

I T was rather odd. Especially when you consider that I had been the editor of PNAS once for five years, and I've been a member of the National Academy of Sciences since 1933. In 1914, the academy decided that members had the right to have their papers published in the *Proceedings*, and this policy of "free access to members" was followed for fifty-eight years. And then—in 1972—they decided to reject my cancer paper. So I ended up submitting the paper to *Oncology*, where it was finally published.[8]

Pauling lost that editorial battle but refused to surrender the war. *PNAS* published his and Cameron's next paper on cancer, in October of 1976.

("We argued with them until we got in!" he would explain, laughing.) Still, he failed year after year to obtain government funding for the cancer-research projects that he formally proposed, which would use vitamin C either with humans or animals.

In 1978 Dr. Cameron took a leave of absence from his position in Vale of Leven Hospital and came to the Linus Pauling Institute, where he and Pauling would complete the book that LPI itself would publish: *Cancer and Vitamin C.* When this book by Ewan Cameron, M.D., and Linus Pauling, Ph.D., appeared in the following year, the long subtitle indicated its ambitious scope: *A Discussion of the Nature, Causes, Prevention, and Treatment of Cancer with Special Reference to the Value of Vitamin C.* The volume was indeed remarkable for its thorough and balanced discussion of the genesis, types, and conventional treatments of cancer that preceded its focus on cancer therapy using ascorbic acid.

The preface briefly introduced the reason why the book was written, the background of the authors' association, and a summary of Cameron's positive clinical findings when using ascorbic acid with patients diagnosed with terminal cancers of various kinds.

SOME years ago we developed the idea that regular high intakes of vitamin C (ascorbic acid, or its several biologically active salts known as ascorbates) play some part both in the prevention of cancer and in the treatment of established cancer.

Evidence steadily accumulates to support this view.

Cancer, of course, is the major unsolved health problem with strong emotional overtones. Although not the major killer, it has become the most feared of all diseases and a major focus of biological research throughout the world. The repeated statement of our views and clinical results in the scientific literature has given rise to much fruitful discussion with colleagues in the scientific and medical fields, and it has also involved us in a massive correspondence with desperate cancer patients seeking advice and help, as well as with their families, friends, and physicians.

For some years we have tried to write personal letters to these patients, family members, friends, and physicians, but meeting this obligation is now beyond our resources. It seems increasingly clear to us that many of these despairing patients lack understanding of (a) the very nature of cancer, (b) the value and the limitations of all conventional (and some unconventional) forms of treatment of cancer, and (c) our own views as to how vitamin C might help them. This book is an attempt to answer these questions.

Cancer is an unpleasant disease. Death by cancer usually involves much more suffering than other ways of death, such as by a heart attack. The cancer patient may lead a life of misery for months or years before his suffering is brought to an end by death. Much of his misery may be caused by the treatment that is given him in the effort to control the disease.

In the United States about 1.9 million people will die this year. About 20 percent of the deaths, 395,000, will be from cancer. Every day about 2,100 people in this country develop cancer and about 1,080 die of cancer. If the incidence and mortality continue at their present rates, one adult in the United States in every three will develop cancer at some time in his life, and one in five will die of the disease.

During the last twenty years about $10 billion has been spent on cancer research, in the effort to get some control of the disease. The budget of the National Cancer Institute for the year 1979 is $900 million and that of the American Cancer Society is $140 million. Despite this great expenditure and the corresponding great effort, not much has been achieved. Some progress has been made in the treatment of some kinds of cancer, especially leukemia and Hodgkin's disease, by new regimes of treatment with high-energy radiation and anticancer drugs. For most kinds of cancer, those involving solid tumors in adults, which lead to 95 percent of the cancer deaths, there has been essentially no change in overall incidence and mortality during recent years.[9]

Whenever *Cancer and Vitamin C* got knocked by medical reviewers, Pauling, to be sure, leapt to its defense, becoming the book's prime defender and publicist. Cameron, never eager to pick fights with his peers in medical practice, was more restrained in any polemics. For instance, he had been careful to present ascorbic acid as an adjunct cancer therapy, not to offer it as a replacement for standard treatments, particularly surgery.

So once more, a book by Pauling written for both the public and health professionals about orthomolecular medicine sent salvos into the barricades of the medical establishment. As the medical authorities continued to be dubious or even vitriolic in their criticism, Pauling seemed to thrive, as always, in the subsequent firestorms returned by physicians' mouthpiece organizations.

In 1979, at a conference called "The Crisis in Scientific Research," Pauling gave a speech detailing his grievances with the government's biomedical-funding agencies. The following year the National Cancer Institute (NCI)—perhaps partly worn down by Pauling's unrelenting

politicizing of the cancer issue among legislators and with the public—
awarded LPI a two-year grant to study vitamin C's effects in preventing
mammary cancer in genetically susceptible mice. It also provided Pauling with an indirect crack at proving the value of vitamin C in treating
human cancer by funding a new clinical study on vitamin C to take
place at the Mayo Clinic—an NCI-approved research site that received
millions of dollars annually in funds for conducting the war on cancer.

The Mayo trial's outcome, as reported in the medical press and then
in the popular media in 1979, made Pauling's belief in the efficacy of
vitamin C with cancer treatment look foolhardy. It seemed far more
misguided than his notorious promotion of that vitamin as an antiviral
remedy. Cancer was such a serious disease that inevitably desperate hope
fastened upon proffered cures. To promise one by easy and inexpensive
means and then fail to deliver smacked of Pied Piperism. Then, too,
there was fierce competition for research funds using various chemotherapeutic agents that had proven effective in animal studies—and the
vitamin-C-megadose regimen was scarcely considered valid by tradition-
alists. Thus the highly publicized failure at the Mayo Clinic in testing
Pauling's belief in vitamin C's restorative powers with cancer was
grouped with prevalent and usually costly traffic in dubious underground cancer medications like Laetrile (except that Pauling would not
profit except in book sales and as an invited speaker). When the Mayo
report appeared, positive findings in previous clinical work, as in Scotland and Japan, and among some daring American individual practitioners, counted as nothing.

This first Mayo Clinic trial purportedly repeated the protocols of cancer studies at Cameron's Vale of Leven Hospital and at the LPI-affiliated
Fukuoka Torikai Hospital in Japan, but the patients had already been
subjected to chemotherapy against Pauling's objections. The second
controlled trial at Mayo was a consequence of Pauling's blistering attack
on the way in which the first trial had been conducted and reported.
However, vitamin C was given for a limited time and then wholly with-
drawn, probably—among other consequences—subjecting the patients
to the notorious "rebound effect," which mimics scurvy symptoms. Both
trials were summarized by him in the chapter called "Organized Medicine and the Vitamins" in his 1986 book, *How to Live Longer and Feel
Better.* Elsewhere, he was apt to be even more incensed about the entire
matter, calling the Mayo Clinic's research fraudulent and unethical. In
any case, a summary of what transpired provides readers with an insight
into the politics of research, particularly regarding funding, reporting,
and prejudgment.

THIS Mayo work has been publicized as refuting the Vale of Leven and Fukuoka Torikai studies. The record shows, however, that the Mayo Clinic doctors did not follow the protocols of these studies. That work has, therefore, only small relevance to the question of how great the value of vitamin C is for cancer patients.

The first Mayo Clinic study showed only a small protective effect of vitamin C. Cameron and I attributed this reported result to the fact that most of the Mayo Clinic patients had already received heavy doses of cytotoxic drugs, which damage the immune system and interfere with the action of vitamin C, and the fact that the controls were also taking vitamin C in much larger amounts than were the controls in Scotland or Japan. Only 4 percent of the Vale of Leven patients had received prior chemotherapy.

In our studies the vitamin C patients took large amounts of the vitamin, without stopping, for the rest of their lives or until the present time, some for as much as fourteen years. In the second Mayo Clinic Study, the vitamin C patients received the vitamin for only a short time (median 2.5 months). None of the vitamin C patients died while taking the vitamin (amount somewhat less than 10 gm per day). They were, however, studied for another two years, during which their survival record was no better than that of the controls, or even somewhat worse. The Moertel paper and a spokesman for the National Cancer Institute, who commented on it, both suppressed the fact that the vitamin C patients were not receiving vitamin C when they died and had not received any for a long time (median 10.5 months). They announced vigorously that this study showed finally and definitely that vitamin C has no value against advanced cancer and recommended that no more studies of vitamin C be made.

Their results provided no basis whatever for this conclusion, because in fact their patients died only after being deprived of the vitamin C. To the extent that their study showed anything, it is that cancer patients should not stop taking their large doses of vitamin C. Yet the study was heralded upon publication as one that reflected adversely on the Cameron-Pauling work.

When this Mayo Clinic paper appeared, 17 January 1985, Cameron and I were angry that Moertel and his Mayo Clinic associates, the spokesman for the National Cancer Institute, and also the editor of the *New England Journal of Medicine* had managed to prevent us from obtaining any information about their results until a few hours before their publication. Six weeks earlier Moertel refused to tell me anything about the work except that their paper was going to be published. In a letter to me he promised that he would arrange for me to have a copy of the paper several days before publication, but he broke that promise.

The misrepresentation by Moertel and his associates and by the National Cancer Institute spokesman has done great harm. Cancer patients have informed us that they are stopping their vitamin C because of the "negative results" reported by the Mayo Clinic.

It is not often that unethical behavior of scientists is reported. Fraud committed by young physicians doing medical research has been turned up several times in the last few years. Improper representation of the results of clinical studies, as in the second Mayo Clinic report, is especially to be condemned because of its effect in increasing the amount of human suffering.[10]

Above all, however, what Cameron and Pauling's cancer work, book, and other messages gave to people with cancer was *hope.* As the new science of psychoneuroimmunology would demonstrate, optimism itself provides a potent orthomolecular drug for the spirit, which may then transfer benefits to the immune system. Crucial to fighting cancer, the immune system may become badly disabled by chemotherapy or radiation, even surgery. In the recovery from any or all of these invasive treatment tactics, vitamin C has proven to be a valuable ally of the body.

Pauling, though remaining ever faithful to vitamin C, became increasingly interested in the value of other micronutrients. Knowing that people would benefit if he discussed the biochemistry of vitamins, minerals, amino acids, and nutrition generally with regard to their action in the human body, in 1986 he wrote yet another popular book: *How to Live Longer and Feel Better.* Partly pieced together from articles he had written or talks he had given in the past, it was full of information both fascinating and readable, and often permitted Pauling's own personality and opinions to come through.

Pauling's new book, again published by W. H. Freeman, brought him further acclaim from the public, which increasingly accepted the idea that vitamins and other dietary supplements could help maintain or improve health, increase longevity, and possibly even prevent or cure disorders, ranging from the common cold and flu to hepatitis and back pain to heart disease and cancer.

In his "Regimen for Better Health" Pauling recommended a specific vitamin-mineral intake, based on Roger Williams's protocol. But he also recommended taking 6 to 18 grams of vitamin C a day. (He himself kept upping his intake with age; at the end of his life he was taking 18 grams.) And he briefly addressed certain lifestyle issues, such as "Keep active; take some exercise. . . . Avoid stress. Work at a job that you like. Be happy with your family." And of course—"DO NOT SMOKE CIGA-RETTES."[11]

In his book Pauling also warned against a common foodstuff that is potentially dangerous when consumed in large amounts. For years he had kept sugar on his hit list, and now he could discuss the subject more fully, knowing that Americans and other nationalities are too fond of sweet stuff.

THE epidemiological evidence that there is a correlation between the amount of cholesterol in the blood, if not in the diet, and the incidence of heart disease is convincing. When the level of cholesterol is decreased, the incidence of coronary disease decreases. The procedure that has been recommended to decrease the level of cholesterol is to cut down the intake of eggs, meat, and other foods that contain cholesterol. The cholesterol ingested in food does not, however, go directly into the bloodstream. It may be that another procedure is even better than reducing the intake of cholesterol. This procedure is to change our intake of nutrients that are known to be involved in the synthesis and destruction of cholesterol in our bodies. Yudkin has convincingly put the sugar sucrose in this category. . . .

It has been shown in a trustworthy clinical study that the ingestion of sucrose leads to an increase in the cholesterol concentration in the blood. This important study was reported by Milton Winitz and his associates in 1964 and 1970. These investigators studied eighteen subjects, who were kept in a locked institution, without access to other food, during the whole period of the study (about six months). After a preliminary period with ordinary food, they were placed on a chemically well-defined small-molecule diet (seventeen amino acids, a little fat, vitamins, essential minerals, and glucose as the only carbohydrate). The only significant physiological change that was found was in the concentration of cholesterol in the blood serum, which decreased rapidly for each of the eighteen subjects. The average concentration in the initial period, on ordinary food, was 227 milligrams per deciliter. After two weeks on the glucose diet it had dropped to 173, and after another two weeks to 160. The diet was then changed by replacing one quarter of the glucose by sucrose, with all of the other dietary constituents kept the same. Within one week the average cholesterol concentration had risen from 160 to 178, and after two more weeks to 208. The sucrose was then replaced by glucose. Within one week the average cholesterol concentration had dropped to 175, and it continued dropping, leveling off at 150, 77 less than the initial value. . . .

This important experiment, in which the only change made was to replace some of the glucose in the diet with sucrose and then return to the sucrose-free diet, shows conclusively that an increased intake of sucrose leads to an increased level of blood cholesterol. Because of the relation between bloodstream cholesterol and heart disease, this experiment ties the consumption of sucrose directly to increased incidence of heart disease. Moreover, the sucrose-cholesterol effect has its biochemical basis established in the fact that fructose, formed in the digestion of sucrose, undergoes reactions in the body leading to acetate, which is then in part converted to cholesterol. This clinical trial conducted by Winitz and his collaborators strongly supports the conclusion reached by Yudkin that sugar (sucrose) is dangerous as well as sweet.

The ordinary diet, with 20 percent of the food energy from sucrose, corresponds to an average intake of 125 gm per day, 100 pounds per year. To cut this intake in half greatly improves the health, decreasing the chance of developing heart disease and other diseases, lowering the blood cholesterol, and strengthening the body's natural defense mechanisms.

You can decrease your intake of sucrose by half very easily by developing some good habits.

1. Keep away from the sugar bowl. Do not add sugar to your tea or coffee. A rounded teaspoonful of sugar weighs 9 gm. Each time that you refrain from adding it to your cup of coffee or tea you decrease your intake of sucrose by that amount.
2. Do not eat prepared breakfast cereals (frosted cereals) with added sugar. Some of these cereals are 50 percent sugar. When you eat a 2-ounce serving you eat 28 gm of sucrose. Eat sugar-free cereals, and add only a small amount of sugar.
3. Do not eat sweet desserts regularly. As Yudkin has pointed out, this does not mean that when you are a guest you should refuse to eat the dessert your hostess has prepared.
4. Do not drink soft drinks (carbonated beverages), except club soda (carbonated water). The usual 6-ounce bottle or can of a cola drink contains 17 gm of sucrose. If you were to drink four of them per day and eat the ordinary American diet, your sucrose intake would be 155 pounds per year, and, according to Yudkin, you would be fifteen times as likely to die of heart disease at an early age as if you restricted your intake to 50 pounds per year by following these rules.

If you keep your intake of sugar down, vitamin C can supply the rest of your insurance against high cholesterol concentration in your bloodstream.[12]

Pauling also pointed out that "vitamin C is involved in the biochemistry of the synthesis and destruction of cholesterol in our bodies"—indicating that regular and sizable ascorbic acid intake should be an important dietary routine for people with family histories of high cholesterol. He also believed it benefited diabetes-prone persons, whose sucrose-metabolizing ability was awry.

How to Live Longer and Feel Better came out when Pauling was eighty-five years old. He was the very picture of healthful, rosy-cheeked senior status—as Joe McNally's close-up cover photo showed him to be. (He was also wearing his now trademark black beret.) Pauling vigorously publicized the book on TV, radio, and speaking tours.

Middle-aged and elderly readers would pay special attention to what he had written about the molecular and physiological mechanisms in growing old.

UNAVOIDABLY, aging is accompanied by the slowing down of the physiological and biochemical processes that go on in the body, by decreasing strength, and by increasing incidence of illness and probability of death. The molecules of deoxyribonucleic acid (DNA) that control the synthesis of enzymes and other proteins undergo changes (somatic mutations) that lead to decreased production of these important substances or to changes in the molecules that decrease their activity. These changes in enzymes throughout the body are compounded by poor nutrition resulting from poor appetite, failure to take supplementary vitamins, and decreased activity of the digestive enzymes. The increase in the number of cells containing chromosomal abnormality contributes to these effects.

One theory of aging is that many molecular changes that build up in the human body with the passage of time are caused by free radicals, atoms or molecules that are especially reactive because they contain an unpaired electron. They can cause changes in the structure and function of important molecules, such as enzymes, and these changes can produce somatic mutations, mutations in the cells of the body. . . .

Every insult to the body, every illness, every stress increases the physiological age of a person and decreases his or her life expectancy. . . . By controlling the common cold, the flu, and other ailments through the intake of supplementary vitamin C and other healthy practices, we not only avoid the discomfort of these diseases but also slow down the rate at which our bodies deteriorate and at which our stores of vitality are used up. Old people and sick people often move rapidly toward death because they do not eat enough food. Their malnutrition is often the result of poverty, but it may also come about because the food does not taste good or smell good to them. The deterioration in the senses of taste and smell may itself be the result of malnutrition, but it is often exacerbated by the toxic products of illnesses, especially cancer, by the changes accompanying the aging process, and by poor health habits, such as constipation.

Good nutrition can decrease the number of these episodes and prevent the onslaught on the physiological age by improving the general health, strengthening the body's natural protective mechanisms, and helping to control illness. To all these ends, the optimum intake of supplementary vitamins contributes heavily.[13]

By the mid-1980s both Cameron and Pauling had become concerned about a serious health problem other than cancer. This one, of recent origin, threatened to reach pandemic proportions if it could not be stopped worldwide. The sexually and blood-transmitted infection could hide for years in an unusually long incubation or latency period, without its host showing significant symptoms. The disease, once it emerged, became known as acquired immune deficiency syndrome—AIDS for short. The proliferation of the human immunodeficiency virus (HIV) —identified by French and American virologists in 1984—ultimately resulted in a collection of opportunistic diseases, some of them previously rare, and any of which brought an AIDS diagnosis. The health of the AIDS sufferer then often rapidly deteriorated, leading to death.

Naturally Pauling and Cameron looked toward vitamin C as a treatment prospect for HIV infection at all stages. From both laboratory studies and clinical results in past experiments, they knew it could be a potent antiviral substance as well as an immune-system booster—important because HIV destroys the T4 helper cells, the chief white-blood-cell warriors in fighting infections. LPI's researchers believed that vitamin C might prevent or at least retard HIV's seemingly inevitable progression to AIDS; it might also ameliorate AIDS-diagnosed patients' vulnerability to additional infections and the notorious wasting syndrome. In 1985 LPI virologist Raxit J. Jariwalla began applying as principal investigator for research grants from different government agencies and research funding foundations. By then the retrovirus had been explicitly identified, and a blood-testing method was developed for the infection through the presence of its antibodies. Not only did the government and foundations turn down the applications, but some of the institute's regular donors, learning of experimental work on

HIV/AIDS going on in LPI's labs, announced that they would stop making further contributions. The public's support of LPI's controversial cancer research had been forthcoming, but the wholesale moralistic condemnation that afflicted AIDS victims now affected the funding of research that hoped to subdue the virus. Pauling—never one to bend to prejudice and narrow-minded dogma—directed that the institute researchers proceed with this work.

Cameron's death from cancer in 1991 put an end to his work on a book about AIDS and vitamin C. Possibly its publication would have publicized the subject well enough to bring in funds from new public and private donors to expand research and set up cooperative clinical trials with physicians. As a small, independent research facility, LPI lacked the clout to make headway in the highly competitive arena of AIDS-research funding. Yet in LPI's largely self-funded series of lab investigations, examining ascorbate's and other micronutrients' promising effects on HIV-infected cell lines, Jariwalla and his colleague Steve Harakeh exchanged information with other researchers and worked at setting up collaborative clinical studies using megadoses of vitamin C orally or intravenously.

Pauling spoke out whenever he could on the crucial importance of conducting vitamin C research with HIV/AIDS, but his appeals for funding from government agencies and private foundations, as well as his letters to medical journals, were mostly dismissed—just as they had been for his cancer studies. In the fall of 1990 LPI's virology team published their first paper regarding *in vitro* suppression of HIV by ascorbic acid, a communication published in *PNAS*. By 1994 AIDS researchers had become discouraged by the lack of progress in devising effective vaccines for the prevention and treatment of HIV infection, in spite of huge efforts and the expenditure of billions of dollars in research. Belatedly, the medical community is considering the use of nutritional tactics for rebuilding a threatened or collapsing immune system—what Cameron and Pauling had futilely proposed almost a decade earlier. Researchers also now discuss the role of oxidative stress in the disease's progression, with the implication that antioxidant vitamins (notably vitamins C, E, and beta-carotene) may considerably reduce the problem. ("It is ironical that the same government agency which turned down our grant applications ten years ago will soon be sponsoring a study on vitamin C in HIV-infected persons," Jariwalla points out. He helped design an NIH clinical protocol.)

Cancer and Vitamin C had proved so useful and was so much in perennial demand by the public that Pauling began preparing a revised, updated, and expanded edition, which was published in 1993. Because of Dr. Cameron's death, he had done the job alone. By then, fourteen years after the book's initial publication, signs were far more favorable for vitamin C's possible importance in preventing or treating cancer. On the basis of statistical data—epidemiological findings and clinical studies—vitamin C appeared also to have value in preventing or treating other conditions, whether minor or transitory, like colds and flu, or life threatening, such as the cardiovascular diseases. Pauling presented some of this new evidence by adding a new preface and expanding the book's appendices.

Especially among people whose personal lives were affected by cancer, Linus Pauling became better known for his fervent advocacy of vitamin C for this disease than for its preventing or treating a cold. Through the years, notes, letters, and phone calls came in daily to him at the institute —to get his advice, to share frustration and pain, or to tell him about a significant success in dealing with cancer of some kind. This continual public response thanking him for his continued concern over alleviating human suffering thus rewarded him for two decades of ridicule and even defamation.

For over twenty years the Linus Pauling Institute has responded to thousands of inquiries by phone and letter for information about using vitamin C, above all in cancer treatment. With limited staff and funds, it cannot accomplish all that is expected of it by health-care practitioners and people with severe, life-threatening afflictions. It has also received hundreds of testimonials that praise Linus Pauling for his courageous promotion of a substance that, from the evidence they present, helped the writer or speaker, a family member, a friend, or a patient in particularly beneficial ways. A significant number of the inquiries and tributes have actually come from members of the medical profession— which says something about contradictions between the official attitude of physicians' organizations and the opinion of individual doctors, whether about cancer prevention and treatment or any other megavitamin tactic in dealing with diseases.

Yet Pauling would be jarred occasionally by a personal incident reminding him that not everyone by any means was receptive to his orthomolecular and megavitamin ideas and that ridicule, scorn, and doubt still prevailed. During the 1980s he recorded a self-note after coming home from a trip. On the airplane he had sat next to a wealthy Texas rancher; clearly a political conservative, the Texan did not discern his seat mate's identity. Pauling was trying to sustain his end of a casual conversation that kept veering off into difficult directions.

TO change the subject, I said that I had just read a book, by Dr. Eugene Robin, in which he said that it is dangerous to visit a doctor. My companion then said, "Yes, it is dangerous. You might get in with one of those kooks who talk about massive doses of vitamin C."

I was taken by surprise and said, "I am the main kook." He replied by saying, "I think that I have seen you. Weren't you on the *Night of 100 Stars?*" . . .

This episode is interesting in showing that most people think of advocates of a high intake of vitamin C as kooks.[14]

In his waning years, though now mostly watching from afar, Pauling had the satisfaction of seeing his institute move forward into new areas of research in molecular biology—that scientific field he had pioneered a half century earlier. In 1992, in essential financial and management retrenching efforts, Zuckerkandl retired as LPI president and director. Linus Pauling, Jr., M.D., took his father's place as chairman of the board of trustees, and Stephen Lawson—who had been at LPI as a researcher and administrator since 1978—was soon appointed the chief executive officer. Fund-raising was now placed in the capable hands of Stephen Maddox, who became director of development; he had already been in charge of LPI's crucial direct-mail solicitation program for seven years. The overloaded staff was cut by about 50 percent so that the institute's operating expenses could begin to stay safely within the budget; income largely comes from contributions (as is true of most nonprofit organizations).

In the process, LPI narrowed its research back to its original mission, to focus on orthomolecular medicine—limiting its current scope of studies mostly to the biochemistry and molecular biology of micronutrients applied to cancer, toxicology, cardiovascular disease, aspects of aging, viral and immunological diseases, and computational genetics. Vitamin C research inevitably would be involved in many of the projects undertaken. Constance Tsao, who for years had carried out many of Pauling's ascorbate lab-study ideas as well as many of her own, was now in charge of the nutritional biochemistry and cancer research group.

Just at this critical generational-transition time a propitious, wholly unexpected, and sizable bequest—from a thrifty Ohio attorney who had never established a personal connection with the institute beyond being a regular donor of small sums—took LPI's financial status nicely out of the red zone. Although LPI's future location is uncertain—it must leave its present site by 1998 because of a change in the City of Palo Alto's zoning laws—the Linus Pauling Institute now appears healthy enough to move into the twenty-first century, taking its founder's name along with it.

By the last decade of the century, the creative research circle was coming around again to Pauling, who had pioneered in the terrain of molecular biology and learned so much through the years about the connection between structure and function in human molecules, cells, and physiological systems. He had identified the first molecular disease, sickle-cell anemia. There was the promise beckoning of genetic engineering that would someday be able to substitute a normal gene for the abnormal one that caused the hemoglobin distortion—and similarly for hundreds, even thousands, of other molecular and degenerative diseases, known or not yet known. (In 1994 LPI virologists were working on a gene-splicing project to produce endogenous insulin in diabetics.) In the meantime, the orthomolecular approach to specific diseases—genetic, degenerative, infectious—would look for corrective biological substances that might prevent or treat symptoms and causes. Or, conversely, it would study the efficacy of restricting or eliminating particular substances from the diet, which in this condition the patient's metabolism could not properly handle. As the 1990s began, Pauling became especially attracted to new research indicating that vitamin C might be valuable in preventing atherosclerosis. He published several case studies on the effect of amino acid lysine, combined with vitamin C, on easing pain in angina pectoris. This highly promising work continues in LPI's cardiovascular research group.

With health professionals and the public, Pauling kept up his orthomolecular message in innumerable ways, as when he gave this talk at the age of ninety.

EVERY person on earth is essentially in poor health. It's called "ordinary good health," but it ought to be called "ordinary poor health." And because of this small amount of vitamin C in the blood, and because vitamin C is involved in so many of the body's protective mechanisms, we are poorly protected against ravages by disease—diseases of all kinds. To be in true good health requires much more vitamin C than people get in their food—much more than the RDA, which in the United States is 60 mg a day for an adult. Because of the low intake of vitamin C, people suffer much more from the onslaught of disease, people have a shorter period of good health, and people die earlier than they should.

I myself am now taking three hundred times the RDA of vitamin C. I began twenty-five years ago taking fifty times the RDA. I had been taking ordinary vitamin pills for many years already, since 1941, but when I went up to fifty times the 60 mg RDA of vitamin C, I felt better. After some years I went up to one hundred times (6 grams), and then to two hundred times, and then to three hundred times the RDA.[15]

Pauling in his prime had always relished a good scientific or political fight, done on equal intellectual terms. Scientists honored him for his extraordinary record of insights and discoveries, and for his skill as an educator; few would deny that he had been the most influential chemist of the twentieth century.

People now largely recognized Pauling's moral courage in the face of scorn and persecution. Retrospectively, even many of his detractors could now see that he had taken the correct stand in protesting both nuclear testing and the Vietnam War. But he was especially endearingly acknowledged as the steadfast guardian of ordinary people's health, which he believed could be improved by vitamin supplementation. On his ninetieth birthday and subsequent birthdays he received hundreds of letters and cards from well-wishers—individuals, couples, families, and even groups who were passionately and enduringly grateful to him for their own or others' well-being, longevity, or recovery from a condition that had once seemed intractable.

Pauling was diagnosed with cancer in 1991. He underwent several treatment procedures, and for some while he managed to maintain an active schedule of work, travel, and public speaking. He continued to take megadoses of vitamin C. It probably helped to ease pain, though it became clear that this much-cherished substance—Pauling's perennially recommended remedy against the ravages of aging—could no longer fend off the inevitable mortality.

When he became too frail to come often into the institute, Pauling spent most of his time at his home near Big Sur, keeping in touch with the staff and other people by telephone. Written communications were now instantly sent and received via a fax machine tended by the ranch's caretakers, Steve and Sue Rawlings. Though Pauling's visitors had to be restricted, several were allowed to interview him. In his last weeks at least one of his four children—Linda, Crellin, or Linus Jr. (Peter lived in faraway Wales)—was always close to his bedside. All three were with him in his final hours.

Linus Pauling died on August 19, 1994, at the age of ninety-three. His life had been long, happy, and marvelously productive, to the everlasting benefit of both science and society. This was exactly the pathway which, as a young man, he had set out to travel through the world.

Linus Pauling:
A Brief Chronology

1901   Born in Portland, Oregon, on February 28, 1901

1914   Decides to become a chemist someday after watching a friend do chemical experiments involving the transformation of matter

1917   Enters Oregon Agricultural College (OAC) at Corvallis (now Oregon State University)

1919–20 Takes a year off from college to help support his family; teaches quantitative analysis at OAC and becomes interested in understanding chemical bonds

1922   Receives B.S. degree in chemical engineering; in fall becomes graduate student at Caltech and begins research on molecular structure of crystals using X-ray diffraction

1923   Marries Ava Helen Miller and publishes first scientific paper, on structure of molybdenum

1925   Receives Ph.D. in chemistry and mathematical physics
First child, Linus Jr., born (with three more to come)

1926–27 Goes to Europe on Guggenheim Fellowship to study new quantum mechanics physics

1927   Becomes faculty member at Caltech in fall

1929   Interest in biology is stimulated by the formation at Caltech of the biology division and the arrival of T. H. Morgan and other notable geneticists from Columbia University

1930   On a trip to Europe, gets Herman Mark's permission to use his new electron-diffraction technique at Caltech for structural studies of gas molecules, including organic substances
Publishes first book, *The Structure of Line Spectra* (with S. Goudsmit)

1931   First and highly influential paper on nature of chemical bond published (insights began in 1929); he usually said that he considered it his favorite scientific paper
Receives Langmuir Prize (American Chemical Society) as the most promising young chemist

1933   Elected to National Academy of Sciences

1934   Does first biochemistry research, on magnetism and oxygen exchange in hemoglobin

Receives first grant from the Rockefeller Foundation, for biomedical research

1935   Coauthors (with C. Bright Wilson) *Introduction to Quantum Mechanics, with Applications to Chemistry*

1936   Starts investigation of antibodies in immune system and structure of proteins

Elected to membership in the prestigious American Philosophical Society

1937   Appointed chairman of Caltech's division of chemistry and chemical engineering

1938   Visiting Baker Professor at Cornell University for spring semester

1939   Publishes *The Nature of the Chemical Bond, and the Structure of Molecules and Crystals*

1940   Paper (with M. Delbruck) about biological specificity and molecular complementariness

Is diagnosed with glomerulonephritis, an often fatal kidney disease, and is put on an experimental salt-free, low-protein diet (this is his introduction to orthomolecular medicine)

1941–45 Works on wartime projects for U.S. government (for which he is later given Presidential Medal for Merit)

1945–49 Determines that sickle-cell anemia is a disease of hemoglobin molecule: the first molecular disease

1945   Begins delivering lectures to the public about the promise and dangers of nuclear energy shortly after the atomic bombs are dropped on Hiroshima and Nagasaki to end World War II

1946   After joining Einstein's Emergency Committee, begins to inform public about the problems inherent in the development and use of nuclear weapons

1947   First textbook, *General Chemistry,* published by W. H. Freeman

1948   Is Eastman Professor at Oxford University for spring semester

Discovers alpha helix model of polypeptide protein structure but withholds publication

Becomes Foreign Member of the Royal Society of London

1949   As proactive president of American Chemical Society, proposes government and industry support of research in academic institutions; also supports program for socialized medicine

1951   Publishes several papers on the alpha helix and pleated sheets in protein structure

1952   Is refused a passport by the U.S. State Department because his anticommunist statements are "not strong enough"; thereby misses important conference in London on protein structure, at which he probably would have seen, as Crick and Watson did, revelatory X-ray diffraction photographs of DNA structure

1954 Awarded Nobel Prize in Chemistry for research into the nature of the chemical bond

Becomes interested in biochemical factors in mental disease

1955 Signs the Russell-Einstein Manifesto (with ten signatories) and the Mainau Declaration (with fifty-two other Nobel laureates), both calling for an end to all war; these statements led to the inception of the international Pugwash Conferences to address peace and other issues; Pauling often attended them

1956 Begins work on biochemistry of retardation and mental illness, with Ford Foundation grant

Purchases (using the Nobel Prize money) a quarter section of land (160 acres) on California coast, south of Big Sur, for use as a vacation site; he calls it Deer Flat Ranch

1957 With Ava Helen Pauling, circulates among scientists a petition to end nuclear-weapons tests

1958 Presents petition to U.N. with over eleven thousand signatures from forty-nine countries

Writes influential, popular book *No More War!*

Pressured by Caltech trustees because of activism; resigns from administrative posts in division of chemistry and chemical engineering

1959 Spends two weeks with Albert Schweitzer at his clinic in Lambaréné, West Africa

1960 Subpoenaed to testify at Senate subcommittee about test-ban treaty circulated in 1957; by refusing to provide names of assistants he risks imprisonment for contempt of Congress

1961 Paulings help organize peace conference in Oslo to prevent nuclear-weapons dissemination

Publishes molecular theory of general anesthesia (hydrate microcrystal)

1962 Pickets outside White House to protest JFK's resumption of nuclear testing; that evening goes inside for a dinner honoring Nobel laureates

Initiates lawsuit for libel against the *National Review,* the first of a succession of protests over press defamation of character

1963 Awarded Nobel Peace Price (1962) for efforts to halt nuclear tests and promote world peace

1964–67 Leaves Caltech to become research fellow at Center for the Study of Democratic Institutions

1964   Publishes *The Architecture of Molecules* with Roger Hayward
Begins producing papers on spheron theory of structure of atomic nuclei

1965   At Pacem in Terris conference at U.N. presents his operative ethical principle of "minimization of suffering"
Continues work with Emile Zuckerkandl on molecular evolution theories, including the molecular clock

1966   Biochemist Irwin Stone arouses his interest in vitamin C for prevention and therapy in health
Is intrigued with results of experimental megavitamin treatment of schizophrenia in psychiatric patients, conducted by Osmond and Hoffer

1967   Goes to India to give several major lectures, which result in the book *Science and World Problems*

1967–69 Becomes research professor of chemistry at UC San Diego
Proposes two new fields: orthomolecular psychiatry and orthomolecular medicine

1969–74 Professor of chemistry at Stanford University

1970   His *Vitamin C and the Common Cold* becomes a bestseller

1973   Cofounds Institute of Orthomolecular Medicine (which in 1974 becomes Linus Pauling Institute of Science and Medicine)

1975   Awarded National Medal of Science

1976   Publishes expanded book, *Vitamin C, the Common Cold, and the Flu*
Gives Centennial Address of the American Chemical Society: "What We Can Expect for Chemistry in the Next 100 Years"

1979   Publishes *Cancer and Vitamin C* with Ewan Cameron, M.D.

1981   Death of wife, Ava Helen Pauling

1984   Given American Chemical Society's highest award, the Joseph Priestley Medal

1986   *How to Live Longer and Feel Better* is published
Announces that his alma mater Oregon State University will be the recipient of his and Ava Helen's papers as well as his collection of medals and other memorabilia

1989   Receives National Science Foundation's Vannevar Bush Award

1993   Twentieth anniversary of Linus Pauling Institute is celebrated

1994   Dies of cancer on August 19 at his coastal ranch near Big Sur

During his lifetime, and even posthumously, Linus Pauling received many national and international honors and awards, too numerous to list here.

Notes on Sources

ABBREVIATIONS

LP = Linus Pauling
LPI = Linus Pauling Institute of Science and Medicine
OSU = Oregon State University/Ava Helen and Linus Pauling Papers
Caltech = California Institute of Technology
UC Berkeley = University of California, Berkeley
MS = Handwritten manuscript
TS = Typewritten script

CHAPTER 1: STARTING OUT

This first-person narrative is a compilation of statements that Pauling wrote or said at widely different times. Individual fragments are not identified, since many pieces have been blended, sometimes several in the same paragraph. The editor has also inserted transitional words or sentences whenever it was necessary to provide linkage, so as to allow Pauling to talk about his early self without interruption, by providing as many of his recollections as possible.

The sources of this chapter are given in chronological order in terms of publication or interview session. Some of them are referenced as sources in subsequent chapters.

The editor thanks the editors, interviewers, organizations, and publishers for permission to adapt their materials in this form.

1. Ridgway, David. "Interview with Linus Pauling." *Journal of Chemical Education*, August 1976.
2. "Molecular Architecture and Molecular Progress." In *The Scientists Speak*, edited by Warren Weaver. New York: Boni & Gaer, 1947. (Based on a radio talk show given by LP in October 1946.)
3. Fry, Ilona J. (oral historian). LP-corrected TS of interview with LP, May 20, 1980. Oregon State History Archives, Oregon State University, Corvallis.

4. "Chemistry." In *The Joys of Research*, edited by Walter Shropshire, Jr. Washington, D.C.: Smithsonian Institution Press, 1981. (Talk presented at a colloquium celebrating the centennial of the birth of Albert Einstein, March 1979.)

5. Lecture on Joseph Priestley for American Chemical Society's 1983 celebration of the English chemist's 250th birthday. MS.

6. Sturchio, Jeffrey. LP-edited TS of unpublished interview with LP, April 6, 1987. Courtesy of the Center for the History of Chemistry.

7. LP-edited TS of filmed interview with LP, November 11, 1990. Courtesy of Wayne Reynolds, American Academy of Achievement.

8. "How I Developed an Interest in the Question of the Nature of Life." First (and only) chapter in an unpublished and unedited memoir drafted in 1992. LPI files.

9. "An Extraordinary Life: An Autobiographical Ramble." In *Creativity: Paradoxes and Reflections*, edited by Harry A. Wilmer. Wilmette, Ill.: Chiron Publications, 1991.

10. Marshall B. Stearn. In *Portraits of Passion. Aging: Defying the Myth.* Sausalito, Cal.: Park West Publishing Co., 1991.

11. Gage, Richard L., ed. and trans. *A Lifelong Quest for Peace: A Dialogue Between Daisaku Ikeda and Linus Pauling.* Boston: Jones and Bartlett Publishers, 1992.

12. Interview with LP conducted by telephone and correspondence. *Elitely Veery* magazine, 1993.

13. Evans, Barry, ed. "A Conversation with Linus Pauling." *Everyday Wonders: Encounters with the Astonishing World Around Us.* Chicago: Contemporary Books, 1993. (Reprinted from the *Journal of Education*.)

14. "Interviews with the Clay Scientists: Linus Pauling." *CMS News*, Spring 1994. Boulder, Colo.: The Clay Mineral Society, 1994.

Laboratory incident on page 36 is taken from a Pauling lecture, "The Advancement of Knowledge," in *Centennial Lectures, 1968–1969* (Corvallis: Oregon State University Press, 1969). The discussion of metals was adapted from several places in Pauling's *College Chemistry*, 3rd ed. (see note 3 for Chapter 2).

CHAPTER 2: WHAT IS CHEMISTRY?

Epigraph from LP's opening words at first Hitchcock Lecture, January 17, 1983. See note 5 below.

1. From the earliest version of *General Chemistry* (a paperbound manual printed by the California Institute of Technology in 1941), 1.

2. *General Chemistry: An Introduction to Descriptive Chemistry and Mod-*

ern Chemical Theory (San Francisco: W. H. Freeman & Co., 1947, 1949), 3–6.

3. *College Chemistry*, 3rd ed. (San Francisco: W. H. Freeman & Co., 1964), 7.
4. Ibid., 9–11.
5. LP's MS for first Hitchcock Lecture, January 17, 1983, UC Berkeley. Although this lecture, "The Development of the Concept of the Chemical Bond," and the second lecture, February 20, 1983 ("Chemical Bonds in Biology"), have not been published, they were videotaped. Handwritten and typewritten texts and notes for this lecture are in the Ava Helen and Linus Pauling Papers at OSU.
6. *College Chemistry*, 87.
7. Ibid., 91.
8. Ibid., 20.
9. *General Chemistry*, 20–21.
10. *College Chemistry*, 25, 24.
11. Ibid., 2.
12. Ibid., 25.
13. "Kekulé and the Chemical Bond," in *Theoretical Organic Chemistry* (London: Butterworths Scientific Publications, 1959).
14. Unpublished LP memoir, item 11 in notes for Chapter 1.

CHAPTER 3: EDUCATION IN SCIENCE

Epigraph from TS of senior class oration at Oregon Agricultural College, May 31, 1922. Courtesy of Special Collections (A. H. and L. Pauling Papers), Kerr Library.

1. "The Significance of Chemistry to Man in the Modern World," *Engineering and Science* 14, no. 4 (1951): 10–12, 14. (Monthly magazine is published by the California Institute of Technology.)
2. "Academic Research as a Career," *Chemical and Engineering News* 28 (1950): 3970–71. (Monthly magazine is published by the American Chemical Society.)
3. "Advice to Students," *Engineering and Science* 18 (1955): 17.

CHAPTER 4: PROBING THE CHEMICAL BOND

Epigraph from American Academy of Achievement interview.

1. "Fifty Years of Physical Chemistry in the California Institute of Technology," *Annual Review of Physical Chemistry* 16 (1965): 1.

2. "Crystallography," *Dictionary of Science and Technology* (San Diego: Academic Press, 1992), 559.

3. LP's unpublished memoir; see item 8 in notes for Chapter 1.

4. LP contribution to "Looking Back on Books and Other Guides," *Physics Today* 34, no. 11 (1981): 249.

5. "An Extraordinary Life: An Autobiographical Ramble"; see item 9 in notes for Chapter 1.

6. See note 1 above.

7. See note 3 above.

8. LP's memoir "Arthur Amos Noyes," *Proceedings of the Robert A. Welch Foundation Conferences on Chemical Research* 20, American Chemistry-Bicentennial, November 8–10, 1976, Houston (Houston: Robert A. Welch Foundation), 101.

9. "My First Five Years in Science": combines text from LP's MS and edited form published in *Nature* 371 (September 1, 1994): 10. TS at LPI.

10. See note 1 above.

11. "Why Modern Chemistry Is Quantum Chemistry," *New Scientist* 108 (November 7, 1985): 54–55.

12. American Academy of Achievement interview; see item 7 in notes for Chapter 1.

13. "The Challenge of Scientific Discovery," 29th William Henry Snyder Lecture (Los Angeles: The College Press, 1967). (From LP lecture, May 17, 1967, at Los Angeles City College.)

14. "The Discovery of the Superoxide Radical," *Trends in Biochemical Sciences*, November 1979.

15. See item 7 in notes for Chapter 1.

16. "Imagination in Science," *Tomorrow* (December 1943), 38–39.

17. MS for first George Fisher Baker Lecture, Cornell University, November 12, 1937.

18. *General Chemistry* (1947), 21 n.

19. See note 17 above.

20. *The Nature of the Chemical Bond and the Structure of Molecules and Crystals* (Ithaca, N.Y.: Cornell University Press, 1939).

21. Modern Structural Chemistry," *Chemical and Engineering News*, July 10, 1946.

CHAPTER 5: MESSAGES IN THE BLOOD

Epigraph from preface to *How Life Began*.

1. Preface to *How Life Began*, by Irving and Ruth Adler (New York: John Day Co., 1957).

2. "The Prospects for Good Nutrition, Better Health and World Peace," in *Nutrition, Health and Peace: LPI Symposia*, vol. 1, edited by Raxit J. Jariwalla and Sandra L. Schwoebel (Palo Alto, Cal.: Linus Pauling Institute of Science and Medicine, 1987, 191–98). (LP talk at symposium given by LPI in 1983; slightly edited here.)

3. "The Discovery of the Alpha Helix," unpublished TS. See note 11, Chapter 6.

4. "How My Interest in Proteins Developed," *Protein Science* 2 (1993): 6.

5. "The Nature of the Bonds Formed by Transition Metals in Bioorganic Compounds and Other Compounds," in *Frontiers of Bioorganic Chemistry and Molecular Biology* (Amsterdam: North Holland Medical Press, 1979).

6. "Chemistry," in *The Joys of Research* (Washington, D.C.: Smithsonian Institution Press, 1981). (Talks presented at colloquium celebrating centennial of birth of Albert Einstein, March 16–17, 1979.)

7. Unpublished LP memoir cited as item 8 in notes for Chapter 1.

8. The second Hitchcock Lecture, "Chemical Bonds in Biology," UC Berkeley. See also note 5, Chapter 2.

9. See note 7 above.

10. See note 3 above; p. 1061.

11. LP's MS for first George Fisher Baker Lecture, Cornell University, November 12, 1937.

12. "One Aspect of the Physical Sciences in Relation to Biology," *Cell Biophysics* 9, nos. 1, 2 (1986). (From symposium in San Remo, Italy, May 5–6, 1983, commemorating 150th birthday of Alfred Nobel.)

13. "The Structure of Proteins," Phi Lambda Upsilon Lecture, Ohio State University, Columbus, Ohio.

14. "Analogies Between Antibodies and Simpler Chemical Substances," *Chemical and Engineering News* 24 (1946): 1064.

15. "Molecular Disease," *Pfizer Spectrum* 6, no. 9 (1958).

16. "The Impact of the Patterson Function," in *Patterson and the Pattersons: Fifty Years of the Patterson Function*, edited by Jenny P. Gousker, Betty K. Patterson, and Miriam Rossi (New York: Oxford University Press, 1987).

17. See note 14 above; pp. 1064–65.

18. See note 12 above.

19. "Molecular Structure and Biological Specificity," in *Excerpts from Classics in Allergy*, edited by Sheldon G. Cohen and Max Samter (Carlsbad, Cal.: Symposia Foundation, 1992), 123.

20. LP statement in *The Meaning of Life: Reflections in Words and Pictures*

on Why We Are Here, edited by David Friend and the Editors of *Life* (Boston: Little, Brown & Co., 1991), 69.

21. "The Place of Chemistry in the Integration of the Sciences," *Main Currents in Modern Thought* 7 (1950): 108–11.

CHAPTER 6: PROTEINS REVEALED

Epigraph from second Hitchcock Lecture (see note 8, Chapter 5).

1. "How My Interest in Proteins Developed," *Protein Science* 2 (1993): 6.
2. *General Chemistry* (1947), 513.
3. Ibid., 515.
4. Second Hitchcock Lecture. See note 8, Chapter 5.
5. Among assorted MSS and notes found in a box in LP's closet at his Stanford apartment; now in Pauling Papers, OSU.
6. See item 13 in notes for Chapter 4.
7. Ibid.
8. See note 1 above.
9. "Molecular Architecture and Biological Reactions," *Chemical and Engineering News* 24 (1946): 1064–66.
10. "Molecular Architecture and the Processes of Life" (Nottingham, Engl.: Sir Jesse Boot Foundation, 1948). (Twenty-first Sir Jesse Boot Lecture, May 28, 1948.)
11. "The Discovery of the Alpha Helix," unpublished paper written in 1982, on file at LPI.
12. See item 13 in notes for Chapter 1.
13. Second Hitchcock lecture. See note 8, Chapter 5.
14. "Our Hope for the Future," in *Birth Defects*, edited by Morris Fishbein. (Philadelphia: J. B. Lippincott, 1963), 165–67.
15. See note 13 above.
16. "Modern Structural Chemistry," Nobel Lecture, December 11, 1954. Printed in several places, including *Science* 123 (1956).

CHAPTER 7: ATOMIC POLITICS

Epigraph from LP's foreword to *Lamp at Midnight*, a play by Barrie Stavis.

1. "The Immediate Need for Interdemocracy Federal Union and Mr. Streit's Proposed Declaration of Interdependence," MS for talk at Union Now meeting on July 22, 1940, at McKinley Junior High School, Pasadena. Courtesy of Pauling Papers, OSU.
2. "The Month in Focus: Proposed Federal Aid to Research in Science and Medicine," *Engineering and Science Monthly* 8, no. 12 (1945).

3. Ibid.
4. "Our Job Ahead," *Chemical and Engineering News* 27, no. 1 (1949): 9.
5. Ibid.
6. "Chemistry and the World of Today," *Chemical and Engineering News* (September 26, 1949).
7. *General Chemistry* (1947), 584.
8. "Unsolved Problems of Structural Chemistry," *Chemical and Engineering News* 25 (1947): 2970–74.
9. Ibid.
10. "The Path to World Peace," in *The Prevention of Nuclear War* (proceedings of a symposium held at the University of British Columbia, March 5–6, 1983), privately printed by Physicians for Social Responsibility.
11. TS of LP talk "Atomic Energy and World Government," given at the Hollywood-Roosevelt Hotel to members of the Hollywood Independent Citizens Committee of the Arts, Sciences, and Professions on Friday, November 30, 1945. Courtesy of OSU Pauling Papers.
12. "An Episode That Changed My Life," unpublished article in LPI archives, intended for submission to *Reader's Digest*, ca. 1984. In LP MS file at LPI.
13. See note 10 above.
14. Succeeding paragraphs were adapted from Pauling-Ikeda dialogue; see item 11 in notes for Chapter 1; pp. 32–33.
15. Chapter 1, "The End of War," from *No More War!* (New York: Dodd, Mead & Co., 1958), 3–9.
16. Pauling-Ikeda dialogue (see item 11 in notes for Chapter 1).
17. "A Disgraceful Act," *The Nation* 178, no. 18 (1954): 378.
18. "The World Problem and the Hydrogen Bomb," *New Outlook* 7, no. 5 (1954).

CHAPTER 8: THE PERILS OF FALLOUT
Epigraph from "Bomb-Test Appeal to the United Nations," 1958.

1. Benjamin Franklin quotation in *General Chemistry*, 3rd ed. (1957), and *College Chemistry* (1964).
2. *No More War!* Chapter 9, "The Need for International Agreements," 178–80.
3. Ibid., Chapter 3, "Radioactivity and Fallout," 45.
4. Ibid., 42–43.
5. Ibid., 43–45.
6. Ibid., Chapter 5, "Radiation and Disease," 90–94.
7. Ibid., 96.

8. Ibid., Chapter 4, "Radiation and Heredity," 82–83.

9. "The Dead Will Inherit the Earth," *Frontier: The Voice of the New West*, November 1961.

10. *Linus Pauling on Science and Peace (The Nobel Peace Prize Lecture)*, with an Introduction by Gunnar Jahn (New York: Center for the Study of Democratic Institutions, the Fund for the Republic, Inc., 1964). (The lecture was published in various journals and volumes, including translations.) Reprinted with permission from the Norwegian Nobel Institute. Copyright © The Nobel Foundation, Stockholm, 1963.

11. *Hearing Before the Subcommittee to Investigate the Administration of the Internal Security Act and Other Internal Security Laws of the Committee of the Judiciary*. United States Senate, Eighty-sixth Congress, Second Session, June 21, 1960 (Washington, D.C.: U.S. Government Printing Office, 1960), 32–33.

12. "An Extraordinary Life: An Autobiographical Ramble"; see item 9 in notes for Chapter 1.

CHAPTER 9: APOSTLE OF PEACE

Epigraph from "Prospects for Global Environment Protection and World Peace . . . " (See note 13 below.)

1. "The Nature of the Problem," speech given at *Pacem in Terris—Peace on Earth Convocation*, sponsored by the Center for the Study of Democratic Institutions, New York, February 1965. Reprinted in several different publications in 1965, including an edition published by Pocket Books.

2. "The Social Responsibilities of Scientists and Science," *The Science Teacher* 33, no. 5 (1966). Reprinted with permission from NTSA Publications, the National Science Teachers Association, 1840 Wilson Boulevard, Arlington, VA 22201–3000.

3. Ibid.

4. Ibid.

5. "Science and the World of the Future," *Man and His World: The Noranda Lectures/Expo 67* (Toronto: University of Toronto Press, 1968).

6. Unpublished speech given in Palo Alto in 1970. Found in LP's apartment closet.

7. "We Have Survived This Decade!" reprinted in *The Churchman* (October 1970) and *John Milton Magazine*, large-type edition (December 1971).

8. "We Must Throw Off the Yoke of Militarism to Achieve Albert Schweitzer's Goal for the World," extant as TS in LPI archives.

9. Ibid.
10. Excerpted from LP's TS (in LPI files) for prepared statement "What Should Be Our Goals?" printed in *Our Third Century: Directions* (A *Symposium Committee on Government Operations), United States Senate, February 4, 5, and 6, 1976,* Committee Print, Ninety-fourth Congress, Second Session (Washington, D.C.: U.S. Government Printing Office, 1976).
11. "What Can We Expect for Chemistry in the Next One Hundred Years?" *Chemical and Engineering News* (April 19, 1976). (This was the American Chemical Society's Centennial Address, April 5, 1976, New York City.)
12. "The Path to World Peace," in *The Prevention of Nuclear War* (proceedings of a symposium held at the University of British Columbia, March 5–6, 1983), privately printed by Physicians for Social Responsibility.
13. "Prospects for Global Environmental Protection and World Peace As We Approach the Twenty-First Century," speech given at the Second Soka University Pacific Basin Symposium, at Soka University's Los Angeles campus, August 22–24, 1990, printed by Soka University, 1992.
14. "Humanism and Peace," *The Humanist* 21, no. 2 (March–April 1961). (An address to the American Humanist Association, Cleveland, Ohio, March 17, 1961.) Reprinted by permission of the American Humanist Association.

CHAPTER 10: MIND AND MOLECULES

Epigraph from "The New Medicine," *Nutrition Today*, September/October 1972.

1. "The Future of the Crellin Laboratory," *Science* 87, no. 2269 (1938): 563. Reprinted in excerpted form with permission from American Association for the Advancement of Science.
2. "Molecular Disease," *American Journal of Orthopsychiatry* 29, no. 4 (1959): 684.
3. Foreword to *Enzyme and Metabolic Inhibitors*, vol. 1, *General Principles of Inhibition* (New York and London: Academic Press, 1953).
4. "The Future of Enzyme Research," in *Enzymes: Units of Biological Structure and Function* (New York: Academic Press, 1956). (Originally given as Fourth Edsel B. Ford Lecture.)
5. "Molecular Disease," *Pfizer Spectrum*, May 1, 1958.
6. "Our Hope for the Future," in *Birth Defects*, edited by Morris Fishbein (Philadelphia: J. B. Lippincott, 1963).

7. "Science and World Problems," in *Enzymes in Mental Health* (Philadelphia: J. P. Lippincott, 1966).
8. "Chemistry," in *Joys of Research*; see note 6, Chapter 5.
9. "Where Is Science Taking Us? The Research Frontier," *Saturday Review*, March 24, 1956.
10. "Biological Treatment of Mental Illness: Academic Address," in *Biological Treatment of Mental Illness: Proceedings of the Second International Conference of the Manfred Sakel Foundation* (New York: L. C. Page, 1966). (Held October 31–November 3, 1962, at the New York Academy of Medicine). Reprinted by permission of American Philosophical Library.
11. "Molecular Disease"; see note 2 above.
12. "The Genesis of Ideas," in *Specific and Nonspecific Factors in Psychopharmacology: Proceedings of the Third World Congress of Psychiatry* [Montreal, June 4–10, 1961] (New York: Philosophical Library, 1963).
13. See item 8 in notes for Chapter 1.
14. "A Molecular Theory of General Anesthesia," *Science* 134 (1961). Reprinted in excerpted form with permission from the American Association for the Advancement of Science.
15. "Vitamins and Orthomolecular Medicine," in *The New Healers: Healing the Whole Person*, edited by Larry Geis et al. (Berkeley: And/Or Press, 1980).
16. "The Advancement of Knowledge," in *Centennial Lectures, 1968–1969* (Corvallis: Oregon State University Press, 1969), 27.
17. "Medicine in a Rational Society," *Journal of Mt. Sinai Hospital* (1969).
18. "Vitamin C: The Key to Health," speech given to the Schizophrenia Foundation in 1991, in *Stop the FDA: Save Your Health Freedom*, edited by John Morgenthaler and Steven W. Fowkes (Menlo Park, Cal.: Health Freedom Publications, 1992).
19. "Science and the World of the Future," in *Man and His World: The Noranda Lectures/Expo 67* (Toronto: University of Toronto Press, 1968).
20. "Orthomolecular Somatic and Psychiatric Medicine," *Zeitschrift Vitalstoff-Zivilisationskrankheiten*, January 1968. (Communication to the Thirteenth International Convention on Vital Substances, Nutrition, and the Diseases of Civilization, Luxembourg and Trier, September 18–24, 1967.)
21. "Orthomolecular Psychiatry: Varying the Concentration of Substances Normally Present in the Human Body May Control Mental Disease," *Science* 160 (1968). Reprinted in excerpted form with permission from the American Association for the Advancement of Science.
22. "The New Medicine?" See note on epigraph for this chapter.

CHAPTER 11: VITAMIN CRUSADER

Epigraph: a favorite maxim of LP in his later, orthomolecular years.

1. "An Apparatus for the Quantitative Analysis of Volatile Compounds in Urine," *Journal of Chromatology* 85 (1973). (Article by Arthur B. Robinson et al., including Linus Pauling.)
2. "Orthomolecular Psychiatry," *Schizophrenia* (1971): 3.
3. "My Love Affair with Vitamin C," *Health Care USA* (Fall 1992), 7.
4. "Good Nutrition for the Good Life," *Engineering and Science* 37 (1974): 7–8.
5. "How Vitamin C Improves Health," *LPI Newsletter* 1, no. 1 (1977).
6. "Medicine in a Rational Society," *Journal of Mt. Sinai Hospital* (1969).
7. "Plowboy Interview," conducted by Kas Thomas, *Mother Earth News*, January/February 1978, 18–19.
8. Statement of Linus Pauling at hearings before the Subcommittee on Health of the Committee on Labor and Public Welfare, U.S. Senate, 93rd Congress, Second Session on S. 2801 and 3867, 14 and 22 August, 1974. From LP-prepared TS on file at LPI.
9. "The New Medicine?" *Nutrition Today*, September/October 1972.

CHAPTER 12: THE NUCLEUS OF CONTROVERSY

Epigraph from "Vitamin C: The Key to Health."

1. "Problems Introducing a New Field of Medicine," *Medical Science and the Advancement of Health*, edited by Robert A. Lanza (New York: Praeger Publishing, 1985).
2. "On Vitamin C and Cancer," *Executive Health* 13, no. 4 (1977).
3. "Preventive Nutrition," *Medicine on the Midway* 27 (1972): 18. A publication of the University of Chicago/Medical and Biological Sciences Alumni Association.
4. "Linus Pauling Talks About Vitamin C, Colds, and Cancer," *The Health Foods Communicator*, no. 4 (May/June 1979). (Talk given by LP at the Fourth Annual Meeting of the Orthomolecular Medical Society, March 3, 1979.)
5. "The Crisis in Scientific Research," in *The Crisis in Scientific Research* (conference in Princeton, New Jersey, October 22–23, 1979), proceedings of the general meeting (October 23, 1979). LP's speech was printed in *LPI Newsletter* 1, no. 9 (Fall 1980).
6. See note 4 above.
7. See note 2 above.

8. "Plowboy Interview," conducted by Kas Thomas, *Mother Earth News*, January/February 1978.

9. Original preface to *Cancer and Vitamin C*, by Ewan Cameron, M.D., and Linus Pauling, Ph.D., originally published by the Linus Pauling Institute of Science and Medicine, Palo Alto, Cal., 1979. The updated expanded edition was published by Camino Books, Inc., of Philadelphia, 1993.

10. Chapter 19, "Cancer," in *How to Live Longer and Feel Better* (New York: W. H. Freeman, 1986), 173–74.

11. Ibid., Chapter 2, "A Regimen for Better Health," 9.

12. Ibid., Chapter 6, "Two Eating Problems," 42–43.

13. Ibid., Chapter 24, "Aging: Its Moderation and Delay," 217–18.

14. From LP self-note folder at LPI. Transcribed from audiotape.

15. "Vitamin C: The Key to Health," see note 18 for Chapter 10.

Acknowledgments

I want to thank a number of persons for their assistance during preparation of this book.

First of all, I am grateful to the staff members at the Linus Pauling Institute (LPI), who were informative and helpful, especially Dorothy Bruce Munro, Linus Pauling's secretary-assistant for twenty-one years. She alerted me to the value of particular documents, talked with me for a great many hours about Pauling, and assisted me in securing permissions from publishers for the excerpts selected (initially many more than could ultimately be accommodated in this book). Dr. Zelek Herman, Pauling's research assistant for fourteen years, provided numerous publications, as well as audiotapes and videotapes of Pauling's lectures and interviews. The bibliographical list he prepared was essential to my work. His and Dorothy Munro's efforts to assemble a complete collection of Pauling's published and unpublished writings have been meritorious. The Pauling archives at LPI provided most of the texts for this book—publications, manuscripts, transcripts of interviews, and other materials.

I have appreciated ideas and editorial feedback from Stephen Lawson, Rosemary Babcock, Stephen Maddox, Constance Tsao, and Raxit Jariwalla, who reviewed parts of the manuscript. My thanks go also to the office staff, who assisted in copying and shipping off numerous bundles of manuscripts.

I am also indebted to Linus Pauling's alma mater, Oregon State University in Corvallis—in particular to Ramesh Krishnamurthy of the Ava Helen and Linus Pauling Papers at Kerr Library, who kindly supplied a number of specially requested documents. Thanks also to OSU president Dr. John Byrne, who is well aware of the interactive value of Pauling's collegiate origins and achievements; and to Melvin George, Kerr Library director, and Clifford Mead, curator of Special Collections. My attendance as invited guest at the three-day conference at Oregon State University, February 28–March 2, 1995, commemorating Linus Pauling's birthday, proved useful and stimulating. At "Linus Pauling: A Discourse on the Art of Biography," Pauling family members, colleagues, and biographers, as well as biographers of other scientists, exchanged a multitude of experiences and views. I hope that this assemblage will become an annual event. There is no end to future possibilities in exploration and conversation.

Two of Pauling's children—Linda Pauling Kamb and Linus Pauling, Jr., M.D. (chairman of LPI's board of trustees)—read the manuscript, provided information, and suggested changes. Even more important, they allowed me to read and copy a number of lectures, speeches, and other manuscripts that they had just discovered in a cardboard box in their father's closet while closing up his Stanford apartment. The scripts—some typed but most handwritten, with a few pieced together from previous talks and perhaps attaching current news articles adorned with trenchant Pauling comments rendered with a thick felt-tip pen—came to me vertically folded. Obviously Pauling had put them in his inside coat pocket for convenience. After returning home from a speaking engagement, he would remove the script, deposit it in the box, and probably never look at it again. The layers of folded scripts and newspaper clippings, which sometimes also contained stray restaurant menus, flight schedules, and self-notes, were rather like a fascinating archeological dig, with the lowest level dating back to the mid-sixties. This extraordinary experience, which went on for some days, put me in more direct touch with Pauling's writing and speaking processes and styles than anything else I read during the year of assembling pieces for this book.

Many valuable commentaries and corrections have come from Dr. Verner Schomaker, chemist colleague of Dr. Pauling and notable in his own right as professor and researcher at both the California Institute of Technology and the University of Washington. Dr. Barclay Kamb of Caltech—geology professor, son-in-law of Linus Pauling, and also, rather handily, my brother—has supplied considerable assistance in the science sections and provided valuable insights. Additionally, Professor Robert Paradowski of Rochester Institute of Technology, Dr. Frank Catchpool, Dr. Fred Stitt, Dr. Jerzy Jurka, and Richard Willoughby have shared anecdotes and insights into Pauling's life and character, as have two of his grandsons, Dr. Alexander (Sasha) Kamb and Barclay James Kamb, Esq.

I also wish to thank Nathaniel Sobel, of Sobel Weber Associates in New York, for taking the concept of this book into the publishing world. Certainly I am most grateful to Bob Bender, vice president and senior editor at Simon & Schuster, for first seeing promise in this project and then assisting at all stages of its development into final book form, which included the decisive and deft cutting that was needed. I am greatly indebted too to Bob's capable and genial editorial assistant, Johanna Li, who handled much of our complex bicoastal communications via phone, fax, Federal Express, and the U.S. mail. Copy editor Chuck Antony's careful work on the pruned manuscript is much appreciated.

On pages 307–10 we acknowledge the many publishers who generously granted permission, mostly without fees, for us to publish excerpts from Linus Pauling's writings, lectures, and interviews that appeared originally in print in their books or periodicals. I am thankful for permission to quote interviews with Pauling that were also transcribed but never (or not yet) published. The Notes on Sources detail these derivations. While we provide the requisite permission statements, in a number of instances, as it turned out, the fair-use rule—covering about 5 percent of a total piece—would have permitted quotation without obtaining permission. But we had sought permission prior to cutting and therefore retain the statement supplied by the publisher. Attempts have been made in all cases to secure permission for materials still in copyright or whose copyrights may possibly have been renewed. Sometimes, however, we found that publishers had disappeared and left no trace, not even with the U.S. Copyright Office at the Library of Congress, which took over the diligent search after we failed to locate these firms. Actually, though, all selections contain words written by Linus Pauling himself, who gave us written permission to use them. Sometimes he was not even paid when his words were put into print, but he was pleased that they were published—just as, I hope, he would be pleased to have some of them here.

Permissions

We are grateful to the publishers, institutions, organizations, and individuals listed below for granting permission to reprint or to publish for the first time selected excerpts from Linus Pauling's writings or recorded words, most of which have been copyrighted. More detailed information about all of these publications and talks by Linus Pauling, as well as others by him, will be found in the Notes on Sources.

Oregon State History Archives, Oregon State University, Corvallis, for Ilona J. Fry's interview with Linus Pauling on May 20, 1980.

Smithsonian Institution Press, Washington, D.C., for "Chemistry," reprinted from *The Joys of Research* by Walter Shropshire, Jr., by permission of the publisher. Copyright 1982.

Chemical Heritage Foundation/The Beckman Center for the History of Chemistry, for Jeffrey L. Sturchio's interview with Linus Pauling on April 6, 1987.

The American Academy of Achievement, for Wayne Reynold's unpublished interview with Linus Pauling on November 11, 1990.

Chiron Publications, for an interview with Linus Pauling by Harry A. Wilmer, published as "An Extraordinary Life: An Autobiographical Ramble," in *Creativity: Paradoxes and Reflections*. Copyright 1991.

Park West Publishing Co., for an interview with Linus Pauling by Marshall B. Stearn, in *Portraits of Passion. Aging: Defying the Myth.* Copyright 1991.

Jones and Bartlett Publishers, for A *Lifelong Quest for Preface: A Dialogue Between Daisaku Ikeda and Linus Pauling*, trans. and ed. Richard L. Gage. Copyright 1992.

W. H. Freeman & Co., for excerpts from *General Chemistry: An Introduction to Descriptive Chemistry and Modern Chemical Theory* and *College Chemistry: An Introductory Textbook of General Chemistry*, originally copyrighted and published in 1947 and 1950, respectively.

California Institute of Technology, publisher of *Engineering and Science* magazine, for several articles: "Molecular Architecture and Biological Reactions" (1946), "The Significance of Chemistry to Man in the Modern World" (1951), "Advice to Students" (1955), "Good Nutrition for the Good Life" (1974).

American Chemical Society, for several articles in its monthly magazine, *Chemical and Engineering News:* "Analogies Between Antibodies and Simpler Chemical Substances" (1946), "Our Job Ahead" (1949), "Chemistry and the World of Today" (1949), "Academic Research as a Career" (1950), "What Can We Expect from Chemistry in the Next 100 Years?" (1976).

Annual Reviews, for "Fifty Years of Physical Chemistry in the California Institute of Technology," reproduced, with permission, from the *Annual Review of Physical Chemistry*, vol. 16. Copyright 1965 by Annual Reviews, Inc.

Academic Press, for several publications: "Crystallography," from *Dictionary of Science and Technology* (copyright 1992); *Enzyme and Metabolic Inhibitors*, vol. 1, *General Principles of Inhibition* (copyright 1953); "The Future of Enzyme Research," in *Enzymes: Units of Biological Structure and Function* (copyright 1956).

The Welch Foundation, Houston, Texas, for "Arthur Amos Noyes," in *American Chemistry—Bicentennial, Proceedings of the Robert A. Welch Foundation Conference on Chemical Research XX*, 1976.

"My First Five Years in Science" reprinted with permission from *Nature* (vol. 371, September 1, 1994). Copyright 1994 by Macmillan Magazines.

New Scientist, IPC Magazines Ltd., for "Why Modern Chemistry Is Quantum Chemistry" (no. 108, November 7, 1985).

Cornell University Press, for introduction to *The Nature of the Chemical Bond and the Structure of Molecules and Crystals*. Copyright 1939.

Irving Adler, for permission to reprint Preface to *How Life Began*. Copyright 1956.

"How My Interest in Proteins Developed," in *Protein Science* 2, no. 6 (June 1993), reprinted with the permission of Cambridge University Press.

Pfizer Inc., for "Molecular Disease," in *Pfizer Spectrum*, 1958.

Oxford University Press, England, for "The Impact of the Patterson Function," in *Patterson and the Pattersons: Fifty Years of the Patterson Function*, 1987. By permission of Oxford University Press.

Los Angeles City College, for "The Challenge of Scientific Discovery: The 29th William Henry Snyder Lecture," The College Press, 1967.

The University of Nottingham, for "Molecular Architecture and the Processes of Life," the Twenty-first Sir Jesse Boot Foundation Lecture, 1948.

"A Conversation with Linus Pauling," reprinted from *Everyday Wonders* by Barry Evans. Copyright 1993. Used with permission of Contemporary Books, Two Prudential Plaza, Suite 1200, Chicago, IL 60601-6790.

The Nation, for "A Disgraceful Act" (vol. 178, no. 18, 1954).

Physicians for Global Survival (Canada), for "The Path to World Peace," privately printed by the Physicians for Social Responsibility (B.C. chapter), from the proceedings on *The Prevention of Nuclear War*, held at the University of British Columbia, 1983.

The Norwegian Nobel Institute, for "Science and Peace," the Nobel Peace Prize Lecture. Copyright © The Nobel Foundation, Stockholm, 1963.

"The Social Responsibilities of Scientists and Science," reprinted with permission from NTSA Publications, from *The Science Teacher* (May 1966), National Science Teachers Association, 1840 Wilson Boulevard, Arlington, VA 22201-3000.

The Noranda Inc., Toronto, Canada, for "Science and the World of the Future," in *Man and His World: The Noranda Lectures/Expo 67*, University of Toronto Press, Toronto, Canada, 1968.

Soka University of America, for "Prospects for Global Environmental Protection and World Peace As We Approach the 21st Century," in *The Second Soka University Pacific Basin Symposium* (August 1990). Speech printed by Soka University, 1992.

"Humanism and Peace," in *The Humanist* 21, no. 2 (March-April 1961), reprinted by permission of the American Humanist Association. Copyright 1961.

Science: The Global Weekly of Research. Three articles reprinted in excerpted form with permission from American Association for the Advancement of Science: "The Future of the Crellin Laboratory" (vol. 87, no. 2269, June 24, 1938); "A Molecular Theory of General Anesthesia" (vol. 134, 1961); "Orthomolecular Psychiatry: Varying the Concentrations of Substances Normally Present in the Human Body May Control Mental Disease" (vol. 160, 1968).

American Orthopsychiatric Association, for "Molecular Disease," reprinted, with permission, from the *American Journal of Orthopsychiatry*. Copyright 1959.

Philosophical Library (Regeen Najar, New York), for two articles: "Biological Treatment of Mental Illness: Academic Address," in *Biological Treatment of Mental Illness: Proceedings of the II International Conference of the Manfred Sakel Foundation* (1962). Copyright L. C. Page, 1966; "The Genesis of Ideas," in *Specific and Nonspecific Factors in Psychopharmacology: Proceedings of the Third World Congress of Psychiatry* (1961). Copyright Philosophical Library, 1963.

Index

molybdenite, 68, 69–70
mongolism, 218
monkeys, rhesus, 188
Morgan, Thomas Hunt, 95, 96, 225
Morley, Edward Williams, 84, 111
Mother Earth News, 250
Muller, Hermann, 96
Munro, Dorothy Bruce, 10, 262
mutations, gene, 131, 173, 210, 212
  vitamins and, 247–48
myoglobin, 127

Nagasaki, nuclear bombing of, 143, 147,
  149, 151, 152, 153, 166, 172
Nation, 155
National Academy of Sciences, 256
National Cancer Institute (NCI), 274–75,
  276–77
National Committee for a Sane Nuclear
  Policy, 169
National Defense Research Council, 116
National Institute of Mental Health
  (NIMH), 218, 241, 259
National Institutes of Health (NIH), 142,
  218, 241, 253, 257–58
National Research Council, 71–72
National Science Foundation, 142
Nature, 73, 223
Netherlands, 187
neutrons, 37, 146
New England Journal of Medicine, 276
New Scientist, 75
Newton, Isaac, 51, 82–83, 111
New York Times, 265
niacin (vitamin B₃), 229, 230, 232, 242, 243
niacinamide, 229, 230
nicotinic acid (nicotinamide), 219, 235
Niemann, Carl, 114
Nixon, Richard M., 155, 192
Nobel, Alfred, 172
Nobel Peace Prize, 12, 13, 171, 178, 181,
  183
  speech given for, 171–78, 190
Nobel Prize in Chemistry, 12, 13, 63, 133–
  134, 178, 181
  speech given for, 63–64, 134
Noranda Lectures, 190
Noyes, Arthur Amos, 40, 67, 70, 71–72,
  73, 74, 75
nuclear radiation, 165–75
  birth defects from, 166, 173–74
  disease from, 165–68
  fallout shelters and, 170–71
nuclear structure, 264–65
nuclear testing:
  atmospheric, moratorium on, 169–70
  efforts for ban on, 161–75, 286
  fallout from, 39, 157, 159, 161, 173–74,
    211, 222
  and limited test-ban treaty (1963), 171–
    172, 174, 181

petition circulated against, 173–75, 178–
  180
nuclear warfare:
  deaths from, 170, 175–76, 201
  description of, 201–3
nuclear weapons, 64, 132, 137, 143–54
  dropped on Japan, 143, 147, 149, 151,
    152, 153, 166
  Pauling's efforts against, 144–54, 156–
    160, 193–94, 200–203, 252
nucleoprotein, 121
nutritional medicine, *see* orthomolecular
  medicine
Nutrition Today, 236

obesity, 211
Office of Scientific Research and
  Development (OSRD), 116, 139–40
Oncology, 272
Oppenheimer, J. Robert, 147, 149, 155–
  156, 157
Oregon Agricultural College (Oregon
  State University), 9
  Pauling's papers at, 9
  Pauling's studies at, 9, 33–42, 86, 92,
    113, 200
organic chemistry, 53, 91–106
  see also biochemistry; hemoglobin;
    immunology, immune system;
    orthomolecular medicine; proteins
orthomolecular medicine, 12, 14, 18–19,
  39, 194, 209, 213–23, 229–38, 239,
  245, 260–61
  aging process and, 259, 266, 280–81
  clinic opened for, 262–64
  institute founded for research on, 257–
    258
  mental illness and, 216–23, 229–38, 239,
    242, 243, 245, 248–49, 259
  problems gaining acceptance for, 265–
    266
  quantitative analysis in, 241, 244, 263
  Stanford's attitude toward, 257
  sugar and, 278–80
  use of term, 258
  see also Linus Pauling Institute of
    Science and Medicine; vitamin C;
    vitamins
orthomolecular psychiatry, 234–38, 243
Osmond, Humphry, 229, 232, 235, 245,
  248
Overton (anesthesia researcher), 227
oxidation, antioxidants, 121, 259, 268,
  282
oxygen meters, 116
ozone layer, in nuclear war, 202

Pacem in Terris, 184
Pacific Basin Symposium, Second, 203
paleogenetics, 210
pantothenic acid, 229, 243